W9-CBN-981

Sequel to Suburbia

Urban and Industrial Environments
Series editor: Robert Gottlieb, Henry R. Luce Professor of Urban and Environmental Policy, Occidental College

For a complete list of books published in this series, please see the back of the book.

Sequel to Suburbia

Glimpses of America's Post-Suburban Future

Nicholas A. Phelps

The MIT Press
Cambridge, Massachusetts
London, England

© 2015 Massachusetts Institute of Technology

All rights reserved. No part of this book may be reproduced in any form by any electronic or mechanical means (including photocopying, recording, or information storage and retrieval) without permission in writing from the publisher.

This book was set in Stone Sans and Stone Serif by Toppan Best-set Premedia Limited. Printed and bound in the United States of America.

Library of Congress Cataloging-in-Publication Data is available.

ISBN 978-0-262-02983-4

10 9 8 7 6 5 4 3 2 1

For Elizabeth McArthur Phelps

Contents

Preface

My interest in the politics and planning of suburban transformation took shape in the early 1990s while I was working as a subcontractor consultant to Croydon Council in South London. Croydon had picked up on Joel Garreau's term "edge city" and cleverly sought to use it to market itself not as an edge city but as a city at the edge of London. I began my academic career as an economic geographer, and naturally my initial interests were framed in terms of the economic gravity of suburban economic nodes such as Croydon. However, Croydon's opportunistic and overtly self-promotional use of the term edge city soon redirected my interests toward the politics and planning of the sorts of transformations taking place in, or planned for, suburbs in the United Kingdom and Europe. Some of these initial interests and, in particular, whether Europe had anything to compare to America's edge cities were explored with colleagues in *Post-Suburban Europe: Politics and Planning at the Margins of Europe's Capital Cities,* published in 2006 by Palgrave Macmillan.

This book draws on two related research projects undertaken from 2008 to 2010, which I was able to augment with funding from the Bartlett School of Planning in 2013. I am extremely grateful for this funding, without which the book certainly would not have been possible. The first of these research projects was funded by the United Kingdom Economic and Social Research Council under the grant title "Governing Post-Suburban Growth" during 2008–2010 (RES 062–23–0924). One outcome of this research was the edited collection *International Perspectives on Suburbanization: A Post-Suburban World?* (published in 2011 by Palgrave Macmillan). That research took me to Tysons Corner, one of six rapidly developing nodes at the edges of capital cities that manifested development issues that piqued my research interests. At first glance, Tysons Corner, an unincorporated community in

Fairfax County, Virginia, on the Capital Beltway, did not appear to offer an auspicious vantage point from which to reconsider aspects of existing urban theory. It was new, surely too new to yield any important insights. Or else it was somewhere and something destined to be overtaken by some other urban form as the urban frontier left it behind. However, contrary to what I had expected, my visits to Tysons Corner prompted me to question elements of existing urban theories. In particular, that some sites in Tysons Corner had already been through their third iteration of development surprised me. It got me to thinking about the land development process and whether it was all of one piece. The existence of lively debate and planning activity on how to remake Tysons Corner simply underlined these interests. For me, the Tysons Corner study became a way into what I believed was something of a bigger research agenda concerning what may come after suburbia.

This agenda certainly included questions about the character of settlements that could be considered post-suburban. It also included the question of what post-suburban politics might consist of. Luckily, help was at hand, for the historian Jon Teaford and postmodernist scholars of what might be called the Los Angeles school of urbanism in particular had provided something of an essential starting point in this regard. However, it also prompted grander thoughts on whether it might be possible to better attune the political economy approaches to suburbanization (of David Harvey and Richard Walker) and the city as a growth machine (of Harvey Molotch and John Logan) and the idea of urban political regimes (of Clarence Stein) to different types of settlements across complex metropolitan regions and the evolution of individual settlements over time.

A second research project, funded by the British Academy in 2010 under the title "The Planning and Politics of Edge City Retrofit" (SG100596), therefore sought to take these ideas further, by exploring the retrofitting of edge cities. This seemed like a great idea, in light of what was going on in Tysons Corner at the time and Ellen Dunham-Jones and June Williamson's 2008 publication, *Retrofitting Suburbia* (Wiley). However, finding examples of suburban retrofit that went beyond the level of individual architectural projects proved difficult. Finding edge city retrofits was also difficult, perhaps reflecting the fact that American suburban development had only briefly assumed this format, as Robert E. Lang has argued in his 2003 publication, *Edgeless Cities* (Brookings Institution Press). Eventually, after some scouring of the Internet, I alighted on two other sites to examine alongside Tysons Corner. These were the Kendall Downtown area, the development issues of which had already been quite widely reported, and Schaumburg, Illinois, which, although a place that had marketed itself strongly as an edge city, had not yet grappled seriously with ideas of transit-oriented density.

This book attempts to bring these concerns together. It extends and develops ideas first aired in articles written with Dave Valler and Andy Wood and published in the journals *Environment and Planning A* and *Urban Studies,* and by myself in *Urban Affairs Review.* In reflecting on the early days of this book, then, I owe a debt to a number of people whom I persuaded to come part of the way on what may have seemed a none-too-promising journey. Dave Valler and Andy Wood, Andrew Dowling, Nick Parsons and Dimitris Ballas, and Fulong Wu deserve special mention for sticking with it without a hint of doubt or regret. Sonia Roitman, Oleg Golubchikov, and Amparo Tarazona Vento have also been great friends and collaborators on research and publications closely related to the ideas discussed in this book. Doubtless they do not share all my views and should be spared any of the blame this book might incur. I am grateful to Miles Irving of the UCL Geography Department's cartographic laboratory for the graphic illustrations and for remaining patient with my requests. I am also grateful to Jim Morin and Dover, Kohl and Partners for granting permission to reproduce the images in chapter 5.

I have also benefited greatly from being part of an international network of scholars examining the phenomenon of global suburbanisms, work funded as a Major Collaborative Research Initiative of the Social and Humanities Research Council of Canada. This network has proved to be an enormously intellectually stimulating as well as an enjoyable and sociable group to be a part of. In particular I am grateful to Roger Keil, director of the network, for his boundless energy and for steadfastly supporting my research and writing on suburbanization.

Several kind folk took pity on a Brit researching America. Robert Bruegmann's correspondence helped greatly to shape some of the ideas contained in this book and with some of the Chicago metro area specifics. I would also like to thank Peter Muller, Richard Grant, Miguel Kanai, Robin Bachin, and Elizabeth Plater-Zyberk at the University of Miami for their hospitality, help, and guidance while I was in the field in Miami-Dade County in 2010, and Tim Chapin for his comments and suggestions on a draft of chapter 5, covering the Kendall-Dadeland situation. Fairfax County planners Linda Hollis and Sterling Wheeler and Schaumburg mayor Al Larson were extremely generous with their time and patient with my requests for meetings, while librarian Jane Rozek was a great help regarding the history of Schaumburg. During the period of 2008 to 2014, more than seventy other people gave their valuable time to talk to me in connection with my investigations. Though they are too numerous to mention here, the contributions of all are gratefully acknowledged.

Finally, I am grateful to the editorial and production team at the MIT Press for their interest, commitment, and hard work in bringing this book to fruition. In particular I would like to thank the Urban and Industrial Landscapes series editor Robert Gottlieb, acquisitions editor Beth Clevenger, assistant acquisitions editor Miranda Martin, and manuscript editor Deborah Cantor-Adams for their support for the project.

1

Introduction: From the Modern Suburb to the Post-Suburb of a Second Modernity

The framework of growth, however hastily devised, tends to become the permanent structure. For better or for worse, the American suburb is a remarkable and probably lasting achievement.
—Kenneth Jackson, *Crabgrass Frontier*

In the years after World War II, American people and businesses of all sorts moved out of cities or approached them by way of the housing tracts, malls, and campus industry and office developments that steadily coalesced into a distinct and expansive new ring of outer suburbs around the major cities of the United States. Unlike the muted and scattered urban extensions of European towns and cities, the outer suburbs came to form a nearly contiguous conglomeration or matrix of elements in an outward expansion of America's major cities.[1] However, like the urban extensions of European towns and cities, they were, from the start, more diverse in their origins and complexions than we have come to believe: sometimes home to the sorts of major employers that attracted subsequent residential development, sometimes representing bedroom communities that demanded shopping malls and employment opportunities, sometimes coalescing further around railroad station towns but at other times stuck out instead in vast expanses crisscrossed or bounded by major highways.[2]

Although there are some reasons to dub the outer suburbs a "geography of nowhere," to introduce James Howard Kunstler's term, there is little doubt that they have come to represent a distinct place of residence and way of life, and the site of much new economic opportunity.[3] Indeed, as Richard A. Walker contended some time ago, it is hard to conceive of the success of postwar American capitalism without these new suburbs.[4] They represented a propitious "spatial fix" (or geographical embodiment of) for the peculiar strengths of American capitalism during this time and helped to generate the profits being produced by American businesses of all sorts during this time.[5] They were the embodiment of America's "Fordist" mass-production, mass-consumption economy.[6] As Robert A. Beauregard has written, "Suburbanization provided investment in new construction and the purchase of consumer goods that, along with rising exports to Europe, anchored national prosperity."[7] During the 1950s and 1960s, "government articulated a national interest in central city revitalization, while at the same time promoting massive redistribution of population and capital investment from central cities to suburbs."[8] The benefits of employment decentralization to the outer suburbs clearly accrued to the federal government in terms of national economic performance.[9] It hardly seems credible that the federal government was not aware of the fact, and it clearly was complicit in encouraging a process that brought it such gains.[10] Federal and state investments in major new road infrastructure, incentives in the form of mortgage relief, a "growth machine" politics, and the fledgling, and permissive, planning control of rural counties created something of a tabula rasa for development on unincorporated land. This at times barely limited market for development saw local banks, insurance companies, real estate brokers, and land speculators, developers, and house builders grow into national business entities. The development of outer suburbia became a national business. The business of producing the built environment—what David Harvey terms the "secondary" circuit of capital may yet become America's primary international business.[11]

This was a business that fashioned a peculiarly American sense of modernity and projected it internationally. "For centuries," Beauregard writes, "US cities had never quite been able to overcome the history, urbanity, and civilizing image of European cities.... What the postwar suburbs gave up in cosmopolitanism and intellectual and cultural depth, they more than made up in prosperity, freedom of choice, and opportunity. Living well was the American revenge on its European origins."[12] Thus, he continues, "although suburbanization was not confined to the United States, the mass suburbanization of single-family detached houses, shopping malls, an automobile-dependent life-style, and low-density sprawl was peculiar to it."[13] By today, Dolores Hayden suggests, suburbs have "overwhelmed the centers of cities, creating metropolitan regions largely formed of suburban parts."[14] So much were cities overwhelmed that at the end of the twentieth century it was possible to regard the outer suburbs of America's cities as little short of a new urbanity—the sort of lasting achievement alluded to in this chapter's epigraph.[15]

Or are the outer suburbs instead by now the urbanity of an older, first modernity? As with so many models and concepts in circulation in urban studies, one might argue that the American outer suburbs we have in mind represent a certain vintage of urbanization. They were the ultimate logical expression of the sorts of personal mobility and the attendant organization of land use promised by the motor car as early as the first decade of the twentieth century, an expression of bureaucratic, organized capitalism, and the rational spatial configuration best suited to fighting late modernity's cold war.

For while the outer suburbs are associated with the phenomenal economic success of the American economy during the 1940s through the 1970s, they have also come to exemplify the contradictions inherent in the urbanization of capital in general and American capitalism in particular—contradictions that were built into the suburban matrix itself. Once a spatial fix allowing American capitalism to flourish, suburbia has now become a barrier to further accumulation in the United States.[16] These contradictions, it can be suggested, were latent in the format of an outer suburban development and the web of interests that stimulated such development. Some of the barriers that suburbs themselves now represent to future accumulation are registered in present interests such as transit-oriented development (TOD), "smart growth," the retrofitting or repairing of suburbia, and New Urbanism, each of which has embedded in it an explicit critique of the interests and of a development format that are considered to have produced suburban sprawl.[17]

There is certainly enough in present academic and popular debates to realize that much of the shine has come off the outer suburbs of this first modernity, but is there enough to glimpse the makings of a distinctly new post-suburbanity? This is the question addressed in this book. It is an important question, for while in many respects the outer suburban matrix is peculiar to America, it has a continuing legacy. It has a legacy within America itself since it is a development format that continues to be used and is likely to go on being used for some time into the future, owing to the many coincident interests involved. It is easy to overlook the fact that this legacy will be felt unevenly within America itself as a federation of governmental and regulatory arrangements pertaining to different vintages of urbanization. A one-size-fits-all urban theory and policy, even within the United States, let alone beyond, are unlikely to suffice.[18] Just as important, the American suburban ideal has fostered a legacy that is yet to be born in many other parts of the world. Ominously, it is only since the 1980s and after the end of the "short American century" that the American model of suburban living is being exported in earnest.[19]

A Second Modernity: Glimpses of Post-Suburbia?

For Ulrich Beck, a politics of a second modernity has emerged as a result of the unintended consequences of a first modernity.[20] Modern capitalism produced a set of significant environmental and social side effects. These have been as much a product of the state as of the private sector—after all, the private and public sectors became barely distinguishable in what John Kenneth Galbraith memorably termed the "technostructure" of society in late modernity.[21] Though what people have in mind when they refer to suburban sprawl is something natural or spontaneous, it should be remembered that sprawl has been thoroughly planned.[22] That is, suburban sprawl is as good an example as any of the technostructure and an associated sense of modernity at work. It has been planned, though doubtless it would also be a good example of the unintended effects of planning interventions. The outer suburbs have made their own significant contribution to the sorts of global environmental risks around which, as Beck observes, the politics of a second modernity revolve, since the resource and energy usage associated with the suburban format of development and living and working is hard to ignore. As George A. Gonzalez has written in *Urban Sprawl, Global Warming, and the Empire of Capital*, "While urban sprawl policies of the United States can be credited with fostering global economic growth and stability, urban sprawl also has … significant liabilities: climate change and oil depletion. Both of these liabilities result directly from the fact that urban sprawl is predicated on the profligate utilization of fossil fuels."[23] And while renewed population growth in central cities and attempts to promote more compact forms of development in existing or new suburban downtowns give the illusion of reduced energy consumption, it should be remembered that much of the consumption associated with populations in such places remains dependent on production and logistics infrastructures that have long since been suburban.

The contradictions of this first modernity are literally seen in concrete in the outer suburbs themselves, in the vast concrete and tarmac expanses of parking lots and structures and building setbacks from curvilinear road patterns. They are registered in the separation of land uses and all that this entails in terms of the daily commute, not just back and forth from home to work but also between home and any number of amenities and services, such as schools, sports, entertainment, and health facilities. They are registered in the swaths of single-story manufacturing, warehousing, retailing, and office facilities lining major roadways. They are seen in the low-density housing and its occupants, which present a formidable political barrier to infill and a greater density of residential development. Yet it is precisely such a reworking of the suburban development format that can help deliver viable public transit and local services, along with significant reductions in energy consumption. Such a reworking of the development format even has the potential to address the once barely imagined externalities of suburbia, such as traffic congestion.

The contradictions of outer suburbia—the unintended consequences of the sort of modern corporate and state planning that were imbricated in the production of American capitalism's distinctive spatial fix—have become further exposed by generally rising oil prices and the recent subprime mortgage crisis, but also by demographic and housing preference changes. These contradictions have become so evident that they seem certain to drive some measure of response in terms of the urbanization of suburbs over the next decades. Whereas for some time after World War II, the nation as a whole gained from the movement of business out of cities, the outer suburbs themselves barely benefited at all, while the central cities and the inner suburbs bore the costs. The unfolding contradictions of the outer suburban spatial fix now appear to raise the specter of a nation no longer benefiting at all, even as the costs to most communities across the metropolitan regions continue to make themselves known.[24]

For the historian Jon Teaford, the internal contradictions of the Fordist outer suburbs were apparent as early as the 1950s, prompting a subtle change in both the character of local politics and the sorts of questions raised regarding the appropriate scale of government attending to suburban development.[25] It is these contradictions that prompted the gradual, almost imperceptible transformation of residential suburbs into distinctly post-suburban communities displaying distinctly post-suburban politics. Though Beauregard is less convinced, I provide some limited confirmation of Teaford's dating of the antecedents of post-suburbia later in this book. In any case, a transformation of suburbs and suburban politics was well under way by the 1970s in the guise of an "urbanization of suburbia" and was conspicuous by the 1980s with the rise of the many outer cities and edge cities that had sprung up at the intersections of radial interstate highways and state parkways and the orbital beltways surrounding America's major cities.[26]

As the various public and private interests invested in suburban development grapple with some of these contradictions, it is apparent that the contradictions also represent enormous opportunities. The estimated 6 million acres of land in suburban corridors, which are developed at around a 0.25 floor-to-area ratio (FAR) as a result of being 75 percent devoted to parking, would supply two-thirds of the projected growth in housing needs and three-quarters of employment growth over the period 2010 to 2030.[27] Thus, for Arthur C. Nelson, "America is changing ... it will mature. This is a contrast to the half century after World War II when America became a suburban nation.... As it matures, America will likely become an urban society."[28]

Yet, if the *Zeitgeist* is of a sequel to suburbia waiting to be written by *some* architects, planners, and civil society organizations under the manifestos for a New Urbanism, TOD, smart growth, and the like, that picture is not one received by all. Indeed—and here's the rub—arguably, the majority of citizens, architects, planners, politicians, land speculators, and construction, banking, and insurance companies are happy for the story of suburbia to carry on. The production of suburbia "adds up to an automated system that is sustained by inertia.... There are few incentives to try anything different," Nelson notes, while the consumers of suburban housing themselves are the key and rather implacable opponents to change and those in need of incentivizing.[29]

As Judith K. De Jong has recently projected, the future pattern of urbanization in America is likely to be somewhere in between these two perspectives, not least because of what Emily Talen describes as the already "fragmented sense of what urbanism in America is."[30] Joel Kotkin has argued that "the basic pattern of the future metropolis will be built upon a predominantly suburban matrix dominated by cars, road connections, and construction as is familiar to the denizens of contemporary Los Angeles, Phoenix, and Houston."[31] The suburbs of 2050 America that Kotkin envisions might form a new paradigm that embodies neither suburban sprawl nor the traditional city format but a multipolar process of suburbanization at greater density and evincing a greater degree of self-containment. Heeding Kotkin's call for the need for better suburbs opens up questions surrounding the potential of any urbanization of the suburbs to deliver a new post-suburban urbanity. While the likes of the New Urbanism and interest in TOD and smart growth have begun to gain some purchase in planning and local governmental circles, they coexist with more established thinking regarding the ease, familiarity, utility, profitability, and viability of a suburban density and format of residential and commercial development. That any signs of a distinctive post-suburban future for America can only just be glimpsed through this fog of counterclaims should not be surprising.

The Structure of the Book

Chapters 2, 3, and 4 elaborate a theoretical perspective on post-suburbia and its potential meaning in urban theory and relevance to policy debates. They are followed by three chapters that illustrate these ideas and concerns in three different post-suburban communities. Finally, chapter 8 marks a conclusion.

If America became suburban in the last half of the twentieth century, it may take the twenty-first century for what is now a suburban nation to become more fully urban again. The glimpses of post-suburban America presented here underline the difficulties of effecting the sort of systemic change that would be needed for such a transformation of the suburbs to occur. However, they also provide evidence here and there of something altered in the popular desire for and experience of the suburban way of life, the political will that can exist not just in incorporated communities but also in surrounding major redevelopment opportunities on unincorporated land, and even partial examples of intergovernmental cooperation that hint at the revival of metropolitan regional-scale governance.

To begin, in the next three chapters of this book I set out the significance of the post-suburban question as it emerges in concerns over how to rework suburban space, given the already apparent issues of the long-term economic and environmental sustainability of automobile-oriented suburbs and ongoing expectations among citizens and aspirations among politicians for the rounding out of suburban communities. The vast majority of the population was born in suburbs of different vintages, and much employment exists there, yet the suburbs continue to play a secondary role to the historical city cores as laboratories for political, policy, and even academic experimentation and discourse.

In and of themselves, suburbs are rarely the focal point of academic theory building.[32] Only very recently have the subjects of suburbia and suburbanization submitted to significant revision and the recognition of a greater historical and present variation in suburbs and their complexion than was previously acknowledged.[33] Yet a cohesive field of suburban studies has yet to emerge from the fragmented approaches to understanding the suburbs found in, for example, planning, sociology, architecture and urban design, urban morphology, postmodern urban theory, and urban and historical geography. This book will, I hope, contribute to a better understanding of one distinctive emerging class of settlements and their economic and political dynamics from among the various settlement types—cities, stable affluent residential suburbs, declining industrial suburbs, newly built residential suburbs—found in the largest metro regions.

In chapter 2, I locate post-suburban communities within the broader metropolitan spaces of which they are a part. The metropolitan urban regions of the United States represent increasingly complex settlement patterns that embody specialized local economies or "trading places"—as William Bogart terms them—and a variety of trajectories of growth and decline.[34] Indeed, some question whether terms such as city and suburb are not "zombie" categories as a result of the ever-widening scope of the urbanization process.[35] Distinguishing a class of post-suburban settlements and considering the potential for evolution of settlements from suburbs to post-suburbs is far from an unproblematic exercise. Yet it can be one ingredient in a theoretical and policy appreciation of the variety apparent within the unity of the urbanization process.

Specifically, commentators have been vague about how to define post-suburban communities in geographic terms, with Robert E. Lang's "edgeless cities" sprawling from inner suburban to exurban locations, while Robert Fishman's "technoburbs" and Rob Kling and colleagues' "post-suburbs" have been defined at an urban regional scale.[36] Perhaps as a result, it is at the county and regional scale that commentators see new relations of governance being fashioned to act on and shape this new urbanity.[37] I argue that post-suburban communities and their politics can and should be positioned within wider metropolitan urban systems. That is, post-suburbs take their place among a range of different settlement types across metro regions, and their dynamics are as worthy of study as are, for example, the decline of industrial suburbs or the gentrification of inner cities.

In chapter 3, I suggest that some of the problems of speaking of a post-suburban era are resolved by placing the emergence of post-suburban politics in historical perspective, one that sees fundamental continuities with the previous automobile-oriented suburbanization.[38] Specifically, the suburbs formed part of a Fordist spatial fix, in which state intervention was deeply implicated. However, the contradictions of state interventions tend to magnify over time, so that the unanticipated consequences of suburbanization become a barrier to further accumulation. The emergence of a distinctly post-suburban politics might be seen as one manifestation of what Ulrich Beck regards as a politics of a "second modernity."[39] Beck's analysis emphasizes the politicization of major environmental risks (the side effects) of modernity and the processes of individualization in society associated with the rise of special interest groups and identity politics. Yet the unanticipated effects of state interventions in promoting low-density suburban development can hardly be overstated, especially in light of their significant contribution to environmental risks such as those underlying climate change.

This historical perspective is also, of course, a geographic perspective because of the different *vintages* of urban development that exist in the United States. Just as American urbanization is not reducible to a single Chicago or Los Angeles model, so too no post-suburban sequel to suburbia can be reduced to the California example sketched out in an early use of the term.[40] The newest automobile-oriented suburbs *may* be more amenable to reworking if they exist in metropolitan regions with an older vintage of urban development by virtue of extant public transit and other infrastructure networks—though, as we will see, this also depends on the other specifics of the particular metropolitan context under consideration. New suburbs in new metro regions, such as Kendall in metro Miami-Dade County, may truly embody a "splintering urbanism" and have limited prospects for redevelopment in the foreseeable future.[41]

A string of commentators have spoken of the new urbanity being fashioned in the outer suburbs. However, it is one that is very much in its infancy and has only begun to be depicted and analyzed in academic terms. It is possible to view the retrofitting of suburbia as insubstantial, as a postmodern affectation of developers concerned with creating a sense of place when marketing newly developed residential communities.[42] However, post-suburban politics, when viewed as an emerging response to the side effects of modernist suburbanization, appears to coalesce around possibly more substantial concerns to urbanize suburbia and to "retrofit" or rework suburban spaces. In and around local debates over the need for, and the financial and technical challenges to, reworking suburban space, we see a post-suburban politics being played out. It is one in which the traditional popular and political ideals embodied in suburban living have been adulterated somewhat.[43] These traditional suburban ideals have met with the emerging contradictions of suburbanization itself in a politics that centers on tensions over the pursuit of private accumulation (growth) and conservation of the environment, the pursuit of growth and provision for collective consumption, and the appropriate scale and vehicles for governing any post-suburban landscape.[44]

In chapter 4, then, some of the important political and governmental challenges to reworking suburban space are elaborated under these three main headings. First, I consider the tension between the pursuit of private accumulation (primarily conducted as a means of underpinning the local fiscal position of suburban communities) and conservation of the built and natural environments. From the outset, environmental amenity has been sought as part of the suburban way of life and jealously protected by suburban communities. However, it has been overlain recently with an additional layer of environmental politics born of the commonly felt side effects of modernity. Second, since almost by definition suburbs of all complexions exist as less than cities—that is, as somehow less than urban in terms of the various amenities and services that are consumed collectively—there has existed a politics of collective consumption alongside the licensing of private accumulation of capital. The question of financing and providing for collective consumption needs necessarily expands the range of local politics into the arena of intergovernmental cooperation. Third, the prospects for the reworking of suburban space are crucially dependent on the extent and manner in which any rescaling of the state can address the increasing latitude of the collective consumption and environmental corollaries to private accumulation. It is little accident that much suburban development has existed, at least initially, on unincorporated county land. By the same token, its successful redevelopment may founder on the lack of a government entity dedicated to financing and enforcing planning and infrastructure investment aspirations.

In chapter 4, I therefore draw a distinction between what I term mark I and mark II post-suburban politics. The former was an early and purely locally oriented response by communities to some of the contradictions of their suburban character—namely, what Teaford has described as the adulteration of suburban ideals with pragmatic political and policy responses to the economic realities of providing for a host of *local* collective consumption needs. What I term mark II post-suburban politics is barely in evidence anywhere across the expanses of American suburban communities, which are desperately in need in a meaningful sequel to suburbia; mark II politics is implied in the view that the environmental side effects of the suburban format of development and many of the collective consumption needs of individual suburban communities can only be addressed at a scale larger than the individual community.

In the next three chapters I present three glimpses of post-suburban America based on research I conducted during the period 2008 to 2012 as part of projects funded by the United Kingdom Economic and Social Research Council and the British Academy. These chapters pull together local planning and economic development documents, relevant newspaper articles, and published and unpublished local histories. Together, the chapters also draw on more than seventy face-to-face and telephone interviews with local and state politicians and planners, private-sector architects and consultant planners, and civic, environmental, and business organizations.

Originally the three glimpses of America's post-suburban future offered by Kendall Downtown (in Miami-Dade County, Florida), Tysons Corner (in Fairfax County, Virginia), and Schaumburg (a suburb of Chicago in Cook County, Illinois) were selected as part of the research in an attempt to tell the story of the reworking of edge cities specifically. However, the comparatively dense edge city format of outer suburban development was only briefly popular with developers and is no longer the norm, especially for new commercial development outside central cities.[45] Moreover, it proved hard to identify many actual instances of the active reworking of the suburban space of such edge cities.

Nevertheless, the three localities afforded a reasonable coverage of the variety of America's postwar suburbs—a point underlined recently by Jan Nijman and Tom Clery—and the challenges presented in any sequel to suburbia, as seen in the summary facts provided for each site in table 1.1.[46] They allow reasonable examination of the contrasting geographic scale of the suburban redevelopment challenge, the contrasting administrative

Table 1.1
Summary Characteristics of Kendall Downtown, Tysons Corner, and Schaumburg

	Kendall Downtown	Tysons Corner	Schaumburg
Administrative status	Unincorporated area, Miami-Dade County, Florida	Unincorporated community, Fairfax County, Virginia	Incorporated village (1956), Cook County, Illinois
Land area	1.3 km^2	8.5 km^2	50 km^2
Population	3,800[a]	19,627[b]	75,386[c]
Years of retrofit	1998–2008	2010–present	Future?

Notes: a. This is the 2000 census figure for unincorporated East Kendall. The population is likely to be considerably smaller. b. 2010. c. 2000.

context of initiating and implementing that challenge, and the contrasting vintages of American metropolitan development in which such a challenge will need to be met. As such, discussion of the three sites is arranged to emphasize progressively the scale of the challenge of reshaping suburban America, ending with Schaumburg, which corresponds less to an edge city and more to the expansive edgeless city format that Robert Lang emphasizes as the present of suburban America. The three sites pose rather different sequels to the suburban story. Their stories offer glimpses of past, present, and future post-suburban America.

The first look at America's post-suburban future, in chapter 5, shows an attempt to fashion a new downtown, Kendall Downtown, for the sprawling Kendall suburbs of Miami-Dade County in Florida during the 1990s. If, as Raymond A. Mohl and Gary R. Mormino hold, "Postwar Florida came to embody and in turn radiate the values of American culture: youth, leisure, consumption, mobility, and affluence," then Miami-Dade County's landscape of "sprawl plus" represents something of the physical incarnation of this culture.[47] Somewhat paradoxically, it is in this newest and most centerless of American urban environments that the New Urbanism movement, with its appeals to the urban morphology and architectural styles of the past, has grown up. Even as the Kendall Downtown continues to evolve, it is already part of the past of New Urbanism. It exists as something of an island of success in a sea of a repetitive low-density, automobile-oriented, suburban sprawl. Though New Urbanism has emerged and grown as something of a new planning orthodoxy, it is also, as chapter 5 stresses, an orthodoxy that has some very real political, governmental, and private corporate limits as a result of the weight of traditional suburban-oriented residential preferences and architectural, planning, construction, financial, and political interests in America today. Of the three site studies presented, the Kendall Downtown story is one that perhaps best highlights the tensions between growth and conservation of the natural environment.

In chapter 6, I recount the story of the growth and current replanning and redevelopment of Tysons Corner, Virginia. Tysons (its publicists and developers have informally dropped "Corner") is perhaps the archetypal edge city.[48] While private-sector land speculators and property developers have been instrumental in its growth, it has also been subject to several plans over the years. The latest of these planning exercises recently won the Daniel Burnham Prize from the American Planning Association. It proposes a significant reworking of Tysons Corner's suburban space into a proper downtown. It also represents something of a present-day test case for similar attempts to retrofit the many edge cities across America. Tysons illustrates clearly how the pattern of government—or perhaps more precisely a lack of government—can shape prospects for a sequel to suburbia, since it persists as a city in waiting on unincorporated county land. Nevertheless, it is an even better test case of how economic growth and collective consumption are intimately related. The irony is that a settlement unleashed by federal and state expenditure on roads for private automobile use is now set to be saved by more federal, state, and county expenditure, this time on improvements in mass public transit.

The Village of Schaumburg, Illinois, which is the subject of chapter 7, was in some important respects born as a post-suburban development. Incorporated with a tiny population in the 1950s, it was conceived and planned almost from the outset as a new kind of city, a regional capital for the northwestern suburbs of Chicagoland. Yet its conception as a particular, very diffuse, type of new city also means that the sheer scale and separation of land uses shed light on some of the difficulties of building post-suburban communities from the majority of suburban expanses of America, even in the public-transit-rich, older, and increasingly regionally planned metropolitan context of Chicagoland. It is its suburban modernity as a planned community that poses the biggest problem to the reworking of space in Schaumburg. Schaumburg has benefited from remarkable continuity and stability in political leadership since its incorporation, though important questions remain over how political leaders will be able to engage and take the resident population with them as they continue to shape this expansive and new kind of outer city in function but not in form. Since Schaumburg was conceived as a new kind of city for the outer northwestern suburbs of Chicago, its local political leaders will have to assume a leadership role in the sorts of intergovernmental cooperation needed to deliver the big-ticket items of expenditure for collective consumption, such as improvements in mass public transit, necessary for a transformation of these communities.

Finally, in chapter 8, I draw together some of the key themes and concerns raised in the opening chapters of the book. In particular, I reiterate how the challenge of reworking suburban settlement space is enormously varied because of the different ways that suburban settlements relate geographically and temporally to the metropolitan regions of which they are a part. These challenges will likely necessitate new arrangements among governments at the county but also the regional level. The new post-suburban politics will not be fashioned by a small group of architects, planners, or politicians. Instead, any reworking of suburban space will be a political process in which all will need to be involved. Since suburban living represents a mass preference, the emerging post-suburban politics will have to command the approval of the mass of resident voters.[49] It will need to be seen by investors and developers to stack up in financial terms. In this respect, a number of policy analysts have begun to provide some of the tools for appraising the costs and benefits of sprawl, though these have yet to gain significant purchase on the thinking of politicians and government planners, transportation and economic development staff, and the preferences of citizens.

2
Locating Post-Suburbs in a Metropolitan Context

This right here is America. Truly when you close your eyes and try to resurrect in your mind the country in which you have spent the last four months, you imagine not Washington with its gardens, columns and complete set of memorials; not New York with its skyscrapers, with its poverty and riches; not San Francisco with its steep streets and hanging bridges, not the mountains, the factories, or the canyons, but this intersection of two roads and a gas station against a background of wires and advertising billboards.

—Ilya Ilf and Evgeny Petrov, *Ilf and Petrov's American Road Trip*

The observation in the epigraph, from the Russian writers Ilf and Petrov's road trip in the United States, is set against a picture not dissimilar to that of Tysons Corner at the opening of Joel Garreau's *Edge City*. What is striking is that two observers from outside the United States saw the essence of looming suburban America fully thirty or more years earlier than most insiders, and sixty years earlier than the frontiersman Garreau. Originally published in 1936 in Russian, Ilf and Petrov's account accurately locates much of the essence of postwar suburban America. The framework of automobility continues to unfold and sets the terms of the debate not only for locating and characterizing the newest urban forms but also for delimiting the possibilities for their eventual evolution into something different.

In this chapter I seek to distinguish and *locate* post-suburban communities within the broader metropolitan spaces of which they are a part. The metropolitan urban regions of the United States represent increasingly complex settlement patterns that embody specialized trading places, in the words of William T. Bogart, and a variety of trajectories of growth and decline.[1] Indeed, in light of these divergent social, economic, ideological, and political trajectories of settlements within emerging megalopolitan urban forms, some have questioned whether terms such as city and suburb are not "zombie" categories, and whether there is anything that can be generically defined as suburban space.[2]

In this chapter I argue that there is a value in distinguishing a category of settlements that are post-suburban. To begin, I note that the critical commentary has some ambiguities when defining a post-suburban era. The literature has also been vague when defining post-suburban settlements in geographic terms. Definitions that focus on the distinctive mix of interests and politics that may be apparent in post-suburban settlements prove more convincing, and in chapter 4 I discuss these approaches. Nevertheless, drawing together the historical, geographic, and political aspects of modern urbanization in a single discussion allows us to place a category of post-suburbs among a range of different settlement types across metro regions, for the dynamics of post-suburbs are as worthy of study as are, for example, those of the declining industrial suburbs or the gentrifying inner cities.[3]

The Post-Suburban Question

The difficulty of defining the causal properties of specifically urban or place-based social processes has plagued urban studies and human geography for some time.[4] Yet this "urban question" remains, and another attempt at defining "urban" seems appropriate in an era of planetary urbanization, an era in which not only the majority of the world's population now lives in officially defined urban areas but in which the experience and way of life of a larger majority can be described as urban.[5] In some respects, planetary urbanization obviates specifying urban processes at all, since by definition all social, economic, political, and ideological processes are urban in a world exhibiting planetary urbanization.

Yet in simple empirical terms, if we accept the proposition of planetary urbanization, then the experience of it is surely predominantly *suburban* and not urban. That is, the majority of the world's urban population and the majority of the world's urbanization are effectively suburban. To suspend our theoretical sensibilities for one moment, empirically speaking, the urban question under planetary urbanization recasts itself as a suburban question. Indeed, empirically this suburban question opens itself up, as Richard Harris and Peter J. Larkham indicate, into a notion of suburban that is itself a complex reflection of the great variety of suburbs apparent historically and today.[6] Moreover, the sense of the variety implied in any "suburban question" (of what, if any, specific causality can be attributed to the suburban character of the urbanization process) has heightened only recently as scholarly attention has increasingly focused on suburbs and suburbanization as subjects worthy of greater empirical scrutiny and theoretical discussion.[7] Several possible trajectories in the development of suburban settlements have become apparent, perhaps most notably the decline of industrial suburbs and the rise of suburbs exhibiting different and distinct ethnic complexions. The suburban question born of these sorts of varying suburban trajectories prompts a variety of other questions regarding urban politics, including those that have remained as the residue of the urban question posed by Manuel Castells some time ago—questions concerning the role of urban social movements in the making and remaking of suburban communities, questions centering on the boundaries between a politics of collective consumption and a politics of urban entrepreneurialism.[8]

To this we could add a distinctly post-suburban question hanging over some communities that make up the suburban reality of planetary urbanization. Some suburbs have continued to evolve from ostensibly bedroom communities to more balanced communities with employment and urban amenities. A similar evolution toward a post-suburban state is evident in those communities spawned by the decentralization of major employment activities. Finally, across the United States one can find communities that were *born* post-suburban in the sense that they were planned from the outset as self-contained cities, though in suburban locations and at suburban densities.

To recognize the very real difficulties—the impossibility, perhaps—of defining distinctly urban social, economic, political, and ideological processes is not quite the same thing as saying that one shouldn't seek to elaborate empirically the differential nature of the urban question as it reveals itself across what are by now complex urbanized regions, as a contribution to the development of urban theory. The geographer in me pushes me toward an inductive—or perhaps more accurately dialectical—approach of observing potentially different types of settlements as a prelude to a theoretical synthesis and, returning to the very real difficulty of distinguishing place-specific or urban processes, placing potentially different types of settlements, their form and politics, in the temporal and geographic continuity of the urbanization process. As such, from a methodological point of view, the question arises as to whether the urban question isn't usefully—and perhaps for the purposes of first approximation—unpacked into urban, suburban, and post-suburban questions.

Classifying settlements in this way—and in particular, distinguishing a class of post-suburbs from suburbs and cities—provides a way of thinking through important facets of the urban question and the urbanization process. The categories might be regarded as ideal-typical cases to which real examples approximate. I return to such a classification in the next chapter when specifying in more detail the likely different urban politics to be found in each type of settlement, and in particular some of the distinctive features of a post-suburban politics.

Despite the lack of consensus on what suburbs are or how they can be defined,[9] I take as a starting point for the discussion that follows Harris and Larkham's definition of a suburb as a settlement (1) in a peripheral location relative to a dominant urban center, (2) partly or wholly residential in character, (3) of low density of development, (4) with a distinctive culture or way of life, and (5) exhibiting a separate community identity, often embodied in a local government.[10] It is a composite definition drawn from a body of literature that until recently has remained small and has grown only incrementally.[11] The continued outward expansion of urban areas renders element 1 in this definition quite arbitrary and to some extent poses a problem when one attempts to distinguish a class of suburban, let alone post-suburban, settlements, not least because it has led to enormous terminological proliferation. For example, Robert E. Lang recently added the term "edgeless city" to an inventory of more than forty other terms already existing in the literature.[12]

In the remainder of this chapter I explore the several ways in which scholars have alighted on the theme of post-suburbia and tried to define a post-suburb. I then seek to locate this class of settlements within the dynamic outward expansion of metropolitan space in broad terms.

Post-Suburbia: A Clean Break with Suburbia?

Commentators on suburbs and processes of suburbanization have typically been torn between two broadly contrasting standpoints. On the one hand are those who regard these developments as simply the most recent manifestation of long-standing processes of suburbanization. As Greg Hise argues, "The historical record suggests ... edge cities are not a new phenomenon. We can trace [their] conceptual roots back to Ebenezer Howard's garden city and the planned dispersion of the industrial city."[13] The seemingly paradoxical juxtaposition of the likes of an edge city and a Garden City is something that Robert Bruegmann traces to patterns of suburbanization beginning even in ancient times.[14]

Taking a more modest time horizon of the last two centuries has the virtue of stressing continuities between eras of modernity and post- or reflexive modernity and associated patterns and processes of urbanization.[15] Thus, for Dolores Hayden, edge cities represent the latest form of processes of suburbanization that date back two centuries or more in the United States. As she explains, "For all its size, Tysons Corner is a suburb that has evolved from automotive building types."[16] James O'Connell enumerates nine layers of suburban development, the latter of which encompass postwar automobile suburbs, interstate-associated exurbs and sprawl, and the present smart growth era.[17] Similarly, Richard Walker and Robert D. Lewis see agglomeration economies as temporarily associated with the development of distinct nodes integral to the inevitably polycentric process of urbanization. As they argue, "If once these districts were close enough to the center to be confused for a single manufacturing center, by the turn of the century, urbanization had reached the metropolitan scale. Since at least 1850, the North American city has grown largely through the accretion of new industrial districts at the urban fringe, becoming multi-nodal in the process."[18] While the United States' mass suburbanization was unique, the fact that it also "gave rise to the edge cities without which suburban development would have been stymied" again stresses the continuities between suburbia and what have been regarded as post-suburban developments.[19] In a similar vein, Ruth McManus and Philip J. Ethington have argued that "suburbs should be subjected to a longitudinal analysis, examining their development in the context of metropolises that usually enveloped them within a generation or two of their forming."[20]

There is little to dispute here, though one inadvertent side effect of focusing exclusively on the undoubted unity of the urbanization process may be missing "a single suburban place changing character over time."[21] McManus and Ethington propose a framework that includes embeddedness, modification of the built environment and social fabric, and an interactive ecology—the reconsumption and reproduction of suburban spaces. In my own work I have tried to be careful to use the term suburb in a way that does not preclude the evidence that existing suburbs have continued to grow and transform into more fully post-suburban settlements along certain dimensions, notably in functional and political-ideological terms. This is especially important in a European context because of the longer history of urbanization and a stronger tradition of urban containment, though even in the United States it is quite apparent that "the period of mature suburbs blends with the post-suburban era."[22]

On the other hand, and in contrast, several commentators had earlier pointed to the apparent novelty of urbanization at various junctures, but with growing intensity since the 1960s, and certainly well before Joel Garreau encapsulated some aspects of this novelty in introducing the term edge city.[23] Robert Fishman had begun to highlight what he perceived as a break in long-established trends of suburbanization. Arguing that suburbs had lost their traditional meaning, he suggested that "with the rise of the technoburb, the history of suburbia comes to an end."[24] Similarly, although Garreau's five criteria for defining edge cities and their classification into three generic types are by no means unambiguous, they generally have been taken to refer to developments of comparatively recent origins (from the 1960s), which are distinct from preexisting suburbs.[25] Most recently, the Los Angeles school of postmodern urbanism has perhaps gone the furthest in arguing the case for a clean break with past patterns of urbanization.[26] Here the call has been to reject "any effort to corset emerging urbanisms into existing (but obsolete) analytical containers."[27]

When the term "post-suburbia" has been used, it has been in rather different ways, not all of which highlight its temporal novelty. Nevertheless, several authors have used the term in a way that clearly sets it apart from "suburbia."[28] W. H. Lucy and D. L. Phillips use the term to "refer to a time period which is succeeding the suburban era and which includes several spatial forms, including a sprawling exurban rural pattern which is of much lower density than most suburbs." More specifically, they define this post-suburban era in terms of "inner suburban population loss and relative income decline, suburban employment increase, suburban out commuting reduction, exurban population and income increase and farmland conversion."[29] Jon Teaford's explicit prefixing of the word also appears to indicate a break with past patterns and processes of suburbanization. Finally, Neil Brenner, addressing contemporary urbanization in a review of the newest "metropolitan regionalism" in the United States, indicates a qualitative difference between such developments and earlier processes of Fordist (or mass-produced and mass-consumed) suburbanization.[30]

In all of this—and regardless of on which side of the continuity-discontinuity divide in perspective one favors—there is the problem of what we might term a "temporal disparity," or differences in the pace and timing at which such post-suburban settlement patterns have emerged in different settings.[31] William T. Bogart has aptly described how "we are living in the past." And for all but the keenest observers—such as Ilf and Petrov, quoted in the epigraph to this chapter—today's apparently postmodern urban landscapes forced themselves on our consciousness only by the 1960s at the earliest, despite being implied in transportation innovations of the 1920s.[32] Elements of the mass suburbanization somewhat peculiar to the United States in the immediate post–World War II period nevertheless have been belatedly imitated quite widely from the 1980s on.[33] Something of this sentiment is also captured by J. W. R. Whitehand and C. M. H. Carr when they highlight how the historical inevitability of adaptation of the built environment "makes our own time quite unremarkable," yet at the same time "sequences of change in the urban landscape have been the subject of ... different conceptualizations."[34] Indeed, it is precisely this detailed work on suburban morphology that has prompted something of a reassessment of the idea of distinct "morphological periods" since "period boundaries are likely to be ill-defined both temporally and spatially."[35] Howard W. Dick and Peter James Rimmer argue that, seen in comparative perspective, cities embody a set of elements that are bundled and unbundled in different settings, and that there have been periods when patterns and processes of urbanization in North American and East Asian cities have converged, most notably the present. Such periods of convergence on, and divergence from, elements of Western urbanization apparent in Southeast Asian cities might be taken as one manifestation of this temporal disparity.[36] Evidence of such a temporal disparity also comes from Henning Nuissl and Dieter Rink's observation of the heavy involvement of real estate companies and anonymous investment funds in the production of urban sprawl in eastern Germany and the partial similarities with Fordist-style residential suburbanization in the United States (at a time when most commentators are highlighting the post-Fordist nature of urbanization), and in Marco Bontje and Joachim Burdack's suggestion that "recent development tendencies in European metropolitan regions bear resemblance to Edge City development in several respects" but are "'typically European' variations on the original Edge City model."[37]

With these variations on a prototypical suburban theme in mind, then, "American cities are not so much different from those in other countries as ahead of them, and we might expect cities elsewhere to follow the 'North American' pattern just as soon as they have enough automobiles, highways, and disposable wealth to make it work," as Kenneth Jackson has written.[38] For all the critical academic, policy, and to some extent popular commentary on the limitations of American suburbs, the implications of temporal disparity, of variations in the pace and timing of development, are clear. Regardless of what happens to the American suburban "model" at home, it is likely to be a model that will continue to have profound effects abroad for many years to come. Indeed, its effects are only just beginning to be felt in selected countries of the global south that are experiencing the urban transition for the first time and have rapidly growing middle classes.

Locating Post-Suburbia

As with distinctions between suburbs and cities, distinctions between suburbia and post-suburbia in geographic terms also prove less than conclusive. One irony is that, unlike the temporal dimension discussed above, the difficulty of adequately *locating and bounding* post-suburbia is part of its analytical attraction and potentially a key element of post-suburbia's definition when compared to established notions of cities, suburbs, and rural settlements. Moreover, the difficulty of placing post-suburbia—and here I include defining its appearance, or form and morphology—has important implications for the politics of post-suburbia.

Mark Gottdiener and George Kephart highlight post-suburbia as a "new form of settlement space."[39] In this respect, several commentators have pointed to reversals in the binaries of city-suburb, center-periphery, and established notions of the concentric or radial ordering of space in city regions.[40] Michael J. Dear and Nicholas Dahmann argue that "in modernist urbanism, the impetus for growth and change proceeds outward from the city's central core to its hinterland. But in postmodern urbanism, this logic is precisely reversed," although their discussion is also suggestive of yet more complex spatial arrangements when they stress that "urban space, time and causality have been altered."[41] They provide few clues here as to the implications of these thoughts, though I return to them later in the chapter. Klaus Brake and colleagues clearly separate post-suburbia geographically from preceding suburban developments when they urge that "in order to correctly define what really happens [in] 'post' suburbanization, we should explicitly refer to the kind of urbanization that is taking place beyond the formerly suburbanized area, in the still rural hinterland."[42]

If post-suburbia embodies a "new form of settlement space," this implies the need to reject binary categories. "It may no longer make sense to look at urbanization as divided between the kind that takes place centrally and that which is peripheral."[43] Paralleling Lefebvre's thought of planetary urbanization, Rem Koolhaas has noted that "if the center no longer exists, it follows that there is no longer a periphery either. Now all is city."[44] The post-suburban spatial form can therefore be contrasted with the centrality and spatial fixity of suburbs and even edge cities. Not all of Garreau's edge cities are noncontiguous with the city-regions they surround, while some have developed on the back of existing suburbs. Indeed, one of the defining features of post-suburbia, whether it be a technoburb or an edge or edgeless city, is the difficulty in locating it or demarking its boundaries in relation to what we have understood as the center and even some of the older suburbs of the modern city. Thus, Robert Fishman compares the traditional suburb to what he terms the "technoburb," arguing that the latter is "at first ... impossible to comprehend. It has no clear boundaries."[45] As he notes, "Unlike old cities, these new cities had no recognizable centers or peripheries; within regions that covered thousands of square miles they included formerly urban, suburban, and even rural elements; their only structure came from the patterns and intersections formed by the superhighway growth corridors that created and sustained them."[46] In a similar vein, Lang has written that "edgeless cities" are "not even easy to locate" because they "spread almost imperceptibly throughout metropolitan areas, filling out central cities, occupying much of the space between more concentrated suburban business districts, and ringing the metropolitan areas' built-up periphery."[47] Whereas suburbs form part of, are integrated with, and can be planned as part of the monocentric city-region, post-suburbia is part of heavily urbanized regions in which there is fragmentation or "splintering" (broadly, the privatization and localization) of infrastructure and service provision. Crucially, compared to their suburban counterparts, post-suburban settlements seem likely to be significantly more detached from the spatial hierarchies associated with such fixed infrastructure networks.[48]

Partly as a reflection of the difficulty in placing it, the mixing of land uses—residential, manufacturing, service industries, idle, agricultural, and park land—is a characteristic that several commentators have also associated specifically with post-suburbia. In the 1960s, Jean Gottmann highlighted the mixed land uses that characterized the macroscale aggregation that was "megalopolis," while such mixing of land uses has also been highlighted as a defining feature of the extended metropolitan areas of Southeast Asia, though at an admittedly different scale and of a different character.[49] For Fishman, while the New Towns derived from Ebenezer Howard's Garden City ideal have been the victims of megalopolis formation rather than an adequate response to it, Howard's idea of a marriage between town and country "still represents the best aspirations of the contemporary residents of the new city"; it is just that this marriage is blurred in a collage of urban, rural, and suburban.[50]

Similarly, Rob Kling, Spencer C. Olin, Jr., and Mark Poster explicitly identify this mixing of land uses as one distinguishing feature of post-suburbia, a discrimination reinforced in Fishman's descriptions of the technoburbs of North America and in Thomas Sieverts's *Zwischenstadt* developments—in-between cities—in mainland Europe.[51] For Kraemer, post-suburbanization refers to a process that "deals with a change in the current 'suburbanization' phase away from the concentric radial patterns of earlier decades toward new spatial patterns, which are sometimes labeled a 'patchwork structure.'"[52] Reference to a patchwork structure sits comfortably within Gottdiener and Kephart's and Dear and Flusty's perspectives on the spatial form of contemporary urbanization.[53] The latter label contemporary urbanization "keno capitalism," wherein "the relationship between development in one lot and another is a disjointed unrelated affair, because earlier conventions or urban industrial agglomeration have been displaced by a quasirandom collage of non-contiguous, functionally independent land parcels."[54]

Along with such a mixing of land uses we might expect a mixing of urban morphological elements within post-suburbia. Here again we might sound a note of caution, since early work in the urban morphology tradition highlighted the mixing of land uses in what was referred to as the "fringe belt," close to existing suburbs.[55] However, while post-suburbia may come to incorporate or subsume such fringe belts, it perhaps rarely coincides with them in the sense that post-suburban settlements are zones of active development compared to the stagnation implied in the idea of a fringe belt.

Not only are different generations of suburbs within any national context liable to have rather different morphologies but also the mixing of different morphological elements within suburbia may have accelerated with the forces driving post-suburban growth.[56] As many as nine different morphological elements—some distinctly urban in character, such as grids—were found in the relatively short time span of 1980–2005 in six U.S. cities.[57] The heroic appeals to history now made not only in New Urbanist–inspired developments but also in the vernacular of some recent suburban developments internationally suggest that quintessentially urban morphological elements may feature in some post-suburban spaces. Moreover, the sense of a mixing of different elements within contemporary processes of urbanization may be most profound in those countries such as China that are experiencing a very rapid urban transition in an era of unprecedented international trade and investment openness. The distinctly different vintages of urbanization and their associated forms or morphologies, which took place *sequentially* and which are so visible to even the casual observer in the United States and feature so strongly in the most familiar theories of the spatial and social structure of the city, are happening *simultaneously* in a country like China. In this context, it may be less easy to read the history of urbanization from its geography in distinct and homogenous rings or layers of development.[58]

One curious thing about edge cities is that it is what is conspicuously absent in the definition that is probably most defining about them as settlements—their appearance or form, such that we know one when we *see* one. More specifically, they are to be found at the intersections of major roadways and are consequently built for the car. Much of the economic logic of suburban and edge city development can be accommodated to the generalities of suitably modified bid-rent theory. Nevertheless, until very recently, the lack of comparable state interventions systematically distorting patterns of locational advantage toward automobile accessible locations elsewhere has tended to create a problem for comparative analysis. Terms such as edge city tend to be invested with the form of U.S. urban development specifically—so obscuring points of comparison in terms of the functioning of similar settlements in other continental and national settings.[59] One difficulty, then, is presented by what Ewa Mazierska and Laura Rascaroli term a "dimensional disparity" apparent in comparisons of post-suburbia in different continental and national settings. They argue that "the difference between the North American and European city seems to be [more] one of proportions than of substance; in the USA ... changes have been much more extreme and extensive."[60] There is a case for arguing that this dimensional disparity, along with the temporal disparity noted above, obscures at least some valid points of comparison between post-suburban forms in different settings.

From the preceding discussion, it seems clear that definitions of post-suburbia based on its geography prove less than satisfactory. Nevertheless, the generality of the term post-suburbia, notably the vagueness of its geographic reach, arguably does give it an advantage over more tightly defined and context-specific terms—such as edge city—in a comparative analysis.

Post-Suburban Politics

Ironically, the difficulty of defining and delimiting post-suburbs can be seen as critical to the politics of post-suburbia. Regulation theory and established theories of urban politics tend to refer to distinct and bounded cities or city-regions, with little regard for the extent to which local politics is entangled in a wider set of nonlocal political relations.[61] "Neither does a particularly good job with different scales," as Melvin Feldman has noted.[62] In short, theories of the growth machine and the urban regime reflect an era of modern city-regions in which suburbs and even the emerging post-suburbs remain oriented to the central city by way of patterns of economic interaction, along with transportation, communications, and water and sewer infrastructures. The importance of nonlocal relations has been referenced in the extant literature, most notably in Harvey L. Molotch and John R. Logan's incorporation of the role of nonlocal (i.e., national and international funds and business interests) capital in the growth machine; however, this says little about the nonlocal *governmental* relations and political arenas in which settlements are enmeshed. This is surprising, insofar as post-suburban settlements are at least partially outside and at best only loosely coupled with the administrative, infrastructure, and service hierarchies of the modern unitary city region. The post-suburbs of America are increasingly woven into webs of economic relations that are independent of established cities and their older suburbs.[63]

The difficulty of delimiting post-suburbia implies at the very least the need to embrace nonlocal political relations in theorizing urban politics and, perhaps more profoundly, a sense of the relational nature of this politics. In the case of Los Angeles, Greg Hise notes that the private development interests producing urban sprawl were conscious of their impacts at multiple scales across the metropolitan region.[64] Along similar lines, Gerald Frug argues that the multiple scalar interests of individuals must be recognized in any renewal of local government and politics in the United States.[65] In short, as Douglas Young and Roger Keil note, not only do "the new topologies in urban regions ... call for a new relational politics," they also speak to the need for a more fully relational urban political theory. However, just as "the politics surrounding and constituting in-between infrastructure is in flux and not yet geared toward the relational reality of the in-between city," traditional urban political theory remains rooted in notions of firmly bounded territorial processes.[66]

The failure of traditional models of urban politics to incorporate nonlocal governmental relations is widely recognized. Yet the mutation of suburban into post-suburban ideology and politics has entailed governance at new spatial scales—most notably the county, but also larger pan-county regions in the United States. Such nonlocal governmental relations stem in significant part from the way in which traditional suburban localist conservative and exclusionary ideology have been compromised by the need for economic development and investment for collective consumption and production, which has entailed the leveraging of resources from nonlocal tiers of government.[67] All of this should come as little surprise, for post-suburbia is derived from, and continues to be reshaped by, a framework of investment in infrastructure networks funded by nonlocal—that is, federal and state-level, in the United States—tiers of government.

Furthermore, I suggest that the "in betweenness" of edge cities, edgeless cities, and technoburbs that constitute the post-suburban economy—an economy caught between forces of centrality and dispersal—is paralleled by an in betweenness of political relations.[68] Since the in betweenness of post-suburbia registers as a nexus of primarily automobile-dependent flows, it is unsurprising that a significant substantive focus of the new post-suburban politics is connected with transportation (and other) infrastructure issues. As Young and Keil note, "the forgotten infrastructural politics of the inbetween city implies a de-colonization from the forces that built the glamour zones at both ends of its existence: the urban core and the classical suburb."[69]

The politics of post-suburbia are in between in one additional sense. Garreau adds to his observation that edge cities rarely coincide with existing government jurisdictions by noting the presence of shadow or private governmental forms.[70] "If we can't democratize Tysons Corner ... when so many American downtowns have been superseded by developments of this kind—a vast amount of American life will never be subject to popular participation and control," writes Gerald E. Frug.[71] It is hard to ignore the dual nature of the challenge of fashioning post-suburban communities in America when, on the one hand, many of the edge cities that might become the focus of efforts to create greater density and mass transit connectivity are unincorporated, and on the other hand, those born in post-suburban communities retain the limited local liability of suburban residential communities.

There is a sense, then, in which the in betweenness of post-suburbia, rather like emergent regional governance arrangements in the southeast of England described by John Allen and Allan Cochrane, is made up of contingent assemblages of agencies (including different tiers of government). Not least among these agents is a burgeoning private technocratic stratum in society that, once largely within the public sector and in service of the state and its policies and administrative hierarchies, is now just as often mobilized in the service of highly particularistic interests in society. The question remains as to whether coherent political coalitions can emerge and sustain themselves in the face of such fragmentation of interests, let alone coalesce around a new politics of suburban and post-suburban retrofit (which I discuss below).

This leads to questions of how the relational nature of suburban and post-suburban politics might be understood. It is important here to try to reconcile a purely topological view of networks of relations with imaginaries and practices that adhere to the territorial jurisdictions of elected government and other government agencies.[72] Here Feldman's notion of "spatial structures of regulation," through which different flows (of materials, value, personnel, information, property rights, authority) and means of orchestrating those flows (such as command, exchange, reciprocity, altruism, or custom) coalesce, comes very close to the sort of territorial assemblages alluded to by Allen and Cochrane.[73] We are familiar with the ways in which flows (notably of materials, information, or personnel) coalesce to become to a greater or lesser degree place-bound, by virtue of the extensive literature on the theory of agglomeration and notions of the scalar dependence of business and government.[74] We are perhaps less familiar with the extent to which the means of orchestrating these flows congeal to create the sorts of territorial assemblages of power referenced by Allen and Cochrane. However, it seems certain that something of the geographic fixity implied in "vertical" means of *government* (through command) has been eroded not only by the continual flux of the exchange of private transactions but also by the growing significance of "horizontal" means (such as reciprocity, modeling, and seduction) of orchestrating various flows and what they imply for the mobility of ideas of, and policies and practices within, suburban and post-suburban *governance*.

City, Suburb, and Post-Suburb: Urban Politics and Settlement Evolution

Theories of urban politics cannot be expected to hold for all times and places.[75] Yet existing theories tend to be limited to addressing a single settlement type, usually the city proper or the "urban," as undifferentiated and, curiously, rather unchanging units of analysis. There remain real difficulties in adequately defining cities, let alone suburbs and post-suburbs, not least because of the unity of the capitalist urbanization process.[76] Yet I suggest there remains analytical value to trying to distinguish these different classes of settlement in order to understand not only the composition of increasingly complex metropolitan regions but also the variable economic, social, ethnic, ideological, and political dynamics shaping them, and also, ultimately, in resynthesis, the unity of the capitalist urbanization process itself.

In particular, it is important to link the analysis of urban politics to a consideration of the different types of urban settlement or the evolution of individual settlements over time. I take these thoughts further in the next chapter, where I consider how the important continuities apparent in the urbanization process are marked by contradictions and manifest themselves in the dynamics of different settlement types over time. For now, I simply wish to suggest a broad scheme by which we might locate the emergence of a class of post-suburban settlements alongside cities within metropolitan-wide processes of urbanization.

The Evolution of Settlements within Urbanized Regions

For some time now the urbanization process in the United States has produced systems of settlements and associated interactions between them that exceed the monocentric city-regions or metropolitan areas of modernity to cover much larger, frequently polycentric, heavily urbanized, even "megalopolitan" regions. For some, these heavily urbanized regions mean that "traditional concepts and labels—'city,' 'suburbs,' metropolises—are 'zombie categories.'"[77] This expansive urbanization of capital has generated a burgeoning terminology describing both the character of these heavily urbanized regions and their newest constituent settlements.[78] What is clear is that the likes of historical cities, their suburbs, outer suburbs, edge cities, edgeless cities, and technoburbs, are specialized locales within wider, multinodal metropolitan or megalopolitan systems.[79]

It is to the differences among settlements and their different growth trajectories that we need to attend in order to speculate on and distinguish the "new" post-suburban politics. Table 2.1 offers a very simple scheme in which I present a number of scenarios of settlement evolution within the modern city, late modern city-region, and what might, following Beck and colleagues, be labeled, for simplicity's sake, the city-region of a "second modernity."[80] In contrast to the relatively predictable linear outward patterns of growth apparent in the modern and late modern periods, highly variable settlement dynamics are apparent within the era of second modernity, which in turn underlines the need to rework established theories of urban politics. In particular, the insights of these existing theories can be more delicately linked to considerations of settlement type, settlement evolution, and structural change. This would include the evolution of suburbs, since "as metropolitan areas have sprawled, suburban ways have evolved."[81]

I am less interested here in commenting on the scenarios associated with the modern city, which are generally well captured in the extant literature on the historical relationship of cities to their suburbs, or on suburbs and suburbanization specifically—though it should be noted that these latter bodies of literature have been subject to revision with the recognition of cities' diversity historically.[82] I am also less concerned with aspects of settlement stasis and decline. In the continued growth within postmodern city-regions there is ample evidence of the stasis and decline of suburbs and indeed cities (scenarios 5, 6, and 7 in table 2.1). Stereotypical notions of suburban homogeneity have contributed to a sense of stability in the

Table 2.1
Urban Development Processes and Past and Possible Future Relationships among Settlement Types

Modern City
1. City → suburb
Late Modern City-Region
2. City → suburb → post-suburb
City-Region of Second Modernity
3. Post-suburb → city
4. Growing suburb → post-suburb → city
5. Stable affluent suburb → stable affluent suburb
6. Declining suburb → sub-suburb?
7. Declining city → suburb

residential character of affluent, ostensibly residential suburbs (scenario 5 in table 2.1). Such notions have been undermined in the United States, where divergence in suburban social and ethnic complexion has been apparent for quite some time, ushering in a greater appreciation of the contemporary variety of suburbs.[83] Another scenario includes the pathway by which cities (notably small former industrial centers), in losing most of their economic function and fiscal capacity, might regress to become suburbs of or bedroom communities for nearby buoyant cities (scenario 7). This is a subject worthy of study in itself, since a number of major cities whose economies were based almost exclusively on Fordist industries have shrunk significantly in the past several decades. An important strand of urban research now rightly focuses on such processes of shrinkage, including the implications for suburbs. Finally, contemporary variations in economic performance are suggestive of a case of the somehow sub-suburban futures facing some severely declining older residential and industrial suburbs (scenario 6).[84]

The focus of this book is rather on the variable growth dynamics in city-regions of late and second modernity (scenarios 3 and 4). In particular, these scenarios are suggestive of trajectories that are very different from the predictable linear outward growth of the modern city-region, a pattern outlined in work on postmodern urbanism and signaled in recent calls to take seriously questions of the evolution of individual suburbs over time.[85]

I begin by speculating on some of the differences between suburban and outer or post-suburban growth in the late modern city-region. Arguably the most notable functional difference is the more balanced employment and residential character of post-suburbia.[86] The balancing of economic, residential, and other functions was apparent in the United States as early as the 1950s and is taken by Teaford to be the signature of post-suburbia.[87] The evolution of suburbs into post-suburbs was also described by Louis H. Masotti, who noted that some suburbs had shed their residential-only character and begun to acquire industrial, retail and office land-uses.[88] More recently, "the renewed linkage of work and residence" has provided the single basic principle of the technoburbs, which Fishman sees as signaling the end of the suburban era. Thus, quite soon after their development, many outer suburbs "became a blend of residential neighborhoods, business parks, and commercial strips. The traditional dichotomy between cities and suburbs no longer described reality."[89] The sorts of manufacturing, retail, and office complexes often associated with discrete outer suburban developments such as edge cities actually have a long history. They have grown stealthily—in the form of landscaped malls and office and research campuses—as an acceptable accommodation to suburban ideals of residential amenity to form nothing less than a new urbanity.[90]

Post-suburbia may well embody an employment concentration of national significance in *qualitative* as well as quantitative terms.[91] Accordingly, we can already point to evidence of suburbs evolving into post-suburbs (table 2.1, scenario 2). We can also speculate on the further evolution of the post-suburban settlements visible in late modern city-regions. Alongside the growing economic gravity of post-suburbia we can consider whether such settlements have become more fully urban in other respects. Arguably some of these new settlements have begun to acquire government functions and civic spaces along with a greater density and mixed use of developments and buildings. In some instances these settlements have been the subject of conscious efforts to plan and "retrofit" such elements.[92] If successful, such developments more broadly signal the genesis of cities from something as traditionally un-urban as a suburban office park–based edge city—sequence 4 in table 2.1.

Conclusion

History, even the recent history of suburban development, suggests there are important continuities in processes of urbanization. For this reason it is difficult to speak of a distinctly post-suburban era, at least in the formally defined terms adopted by some. As a corollary to this observation, the origins of present-day post-suburbs are varied; some have evolved from existing, primarily residential suburbs, others may emerge from peripheral employment centers, and yet others have been born post-suburban. Post-suburbia is also something that is difficult to define or locate geographically. Yet, because U.S. metro regions are by now complex and variegated socioeconomic and political landscapes, it is surely worthwhile to try to identify different types of settlements, including post-suburbs. In particular, the most convincing evidence of any emergent post-suburban America is to be glimpsed in the changes that have in some communities and for some time been apparent in the ideology and politics associated with the suburban way of life. It is the politics and ideology associated with these communities that may best mark them and reveal the possibilities for a post-suburban America, for a reworking of suburban space into something distinctly more urban. Here urban theory can seek to do more to acknowledge the variegated political landscape that exists across the largest metropolitan areas, and in particular to begin to define the place of post-suburban settlements and distinguish their politics from that of historical cities and suburbs of all kinds.

3 The Suburbs and Their Contradictions: The Post-Suburban Politics of a Second Modernity

For the past thirty years all sorts of people, most of them white, had been moving in beneath those trees, into all those delightful, leafy, rolling rural communities that surrounded the city proper. By their hundreds of thousands they had come . . . into those subdivided hills and downs and glens and glades beneath the trees. . . . There was Spaghetti Junction, as it was known, where Highways 85 and 285 came together in a tangle of fourteen gigantic curving concrete-and-asphalt ramps and twelve overpasses. . . . And now he could see Perimeter Center, where Georgia 400 crossed 285 . . . now Perimeter Center was the nucleus around which an entire edge city, known by that very name, had grown.

—Tom Wolfe, *A Man in Full*

Defining post-suburbia in terms of a new era of urbanization has proved less than satisfactory. Such definitions show their failings when "post-suburbia" is removed from a broader structural context of fundamental continuities with previous processes of suburbanization. Specifically, many suburbs formed part of America's Fordist mass-production "spatial fix," in which state intervention was deeply implicated. However, these state interventions make their own distinctive contributions to the contradictions of capitalism and urbanization, contradictions that can magnify over time, with the result that the unanticipated consequences of suburbanization come to be a barrier to further accumulation.[1]

It is in this context that we might glimpse a class of post-suburban settlement as it is fashioned from a distinctly new post-suburban politics—a politics that is a concrete, though localized, expression of what Ulrich Beck regards as a more generalized politics of a second modernity in urban development. Beck's analysis emphasizes the politicization of major environmental risks (the side effects) produced by the rational corporate planning and state interventions that have been the hallmark of modernity and by the processes of individualization in society associated with the rise of special interest groups and identity politics.[2] Here, then, the unanticipated effects of state interventions in promoting suburban development can hardly be overstated as an ingredient in the politics of a second modernity, owing to their significant contribution to environmental risks such as global climate change.

Moreover, the emergence of a class of post-suburban settlements—defined largely in terms of ostensibly new post-suburban politics and ideologies—can be seen as a manifestation of local experimentation aimed toward a new, post-Fordist, post-suburban spatial fix. This is true despite the temptation among commentators to speak of American urbanism and planning solutions in the singular. The United States embodies a variety of vintages of urbanization and potential models not only of urban theory but also of urban practice, to the point that the nation might be regarded in many important respects as increasingly *disunited* when it comes to the possibilities and politics surrounding urban form and function. There are important divergences in urban regions' possibilities for fashioning a sequel to suburbia with respect to the scale of the environmental, transportation, and infrastructural obstacles to changing present suburban development formats and instituting new governance arrangements. Thus, while commentators have spoken of the new urbanity being fashioned in the outer suburbs, it is an urbanity that is very much in its infancy and one that has only begun to be depicted and analyzed in any depth. Moreover, this new urbanity is not one that is arriving in one piece. In some instances it has evolved organically from traditional residential suburbanization and the associated politics. In other instances settlements have been born post-suburban—representing from the start regional-scale cities—though often lacking a corresponding regional-scale or (what I later term a mark II post-suburban) political sensibility.

Post-suburban politics, viewed as an emerging response to the side effects of modernist suburbanization, appears to coalesce around the variable will of interested parties to engage with a series of substantial and difficult governance challenges, captured in shorthand in terms like the urbanization or "retrofit" of suburbia. It is at present possible to view the retrofitting of suburbia as insubstantial, as a postmodern affectation of developers concerned with marketing particular localities.[3] However, in and around local debates over the need for, and the financial and technical challenges to, reworking suburban space, we see some significant evidence of a post-suburban politics being played out. It is a politics, though, in which the traditional popular and political ideals embodied in suburban living have been somewhat adulterated.[4] These traditional suburban ideals have run up against concerns over the fiscal capacity of suburban village, city, and county governments and the rounding out of suburban communities in terms of amenities, facilities, and employment opportunities. Moreover, these pragmatic local initiatives have more recently run up against emerging movements and planning principles that have gained some currency nationwide, such as New Urbanism, smart growth, and transit-oriented development (TOD).

Continuity in Change: Urban Political Theory and the Unintended Consequences of State Intervention

In this initial section I suggest that insights from examining the emerging gravity of post-suburbia and its geography can be related to the role of the state in the unfolding contradictions in the urbanization of capital. In exploring the emergence of a distinctly post-suburban politics, then, I seek to make a connection between the logic of the late modernist capitalist state and its interventions.[5] I focus specifically on those interventions that produced postwar suburbanization, but also its latent if inherent contradictions.[6] It is these contradictions of postwar suburbanization that now partly appear to be bound up in a politicization of side effects within a second modernity, not least as a result of the juxtaposition of these contradictions of suburbia with the systemic, machinelike properties of the (sub)urban land development imperative, which in large part has continued unabated. The problem here is that existing theories of urban politics (such as growth machine and urban regime theory) have relatively little to say about either these contradictions of the suburban spatial fix and the role of state intervention in producing them, or the different tiers of government implicated. They can usefully be modified to factor in such considerations.

In the most general terms, as Emily Talen has noted, "the whole act of trying to create a better settlement form can be reduced to something that is merely reflective of the contradictions of western capitalist democracies."[7] Indeed, it has been said that planning, or perhaps state intervention more broadly, stands for the many unresolved contradictions in the organization of society at large.[8] As a result of these contradictions, there is an intimate relationship between planning and the market, to the point that theorists have debated whether the activities of sectoral and spatial planning can be said to have distinct effects from those of the "market".[9] Moreover, the unintended consequences of planning interventions in the urban sphere have been significant enough to prompt the suggestion that "'market failures' may not be any more severe than 'government failures' in the production of sprawl."[10]

More specifically, state intervention arises out of the contradictions and crisis tendencies associated with the accumulation of capital in general and the urbanization of capital in particular. As A. J. Scott and S. T. Roweis write, "State intervention can be understood only as a continual stream of responses to the negative and disruptive outcomes of the unresolved—and in a capitalist society, the unresolvable—contradiction between privatized and decentralized decision making . . . and collective action."[11] The modernity of state interventions of all sorts has been called forth by what Claus Offe terms a "double weakness" of the capitalist accumulation process: the weakness of being unable to produce all of the necessary inputs to accumulation through the accumulation process itself, and a weakness stemming from the fact that the capitalist class, being made up of essentially competitive units, is associated with a fundamentally fragmented politics surrounding questions of how the state should intervene.[12] Thus, it is perhaps forgotten that despite suburban house building and big associated changes in traffic conditions across metropolitan areas prior to the 1950s, business interests remained fragmented, each pursuing its own specific interests in respect to lobbying for road building at different parts of the road network.[13] It could hardly be said that the federal government had intervened decisively in a way to favor suburbanization until this point.

In this way, the contradictions of the urbanization process call forth an endless stream of interventions. These interventions have been viewed very much as a case of planning as "muddling through": of piecemeal, incremental, and local government interventions.[14] With respect to the many incremental interventions by local governments, the successive contradictions produced are now familiar enough in the form of the vagaries of forecasting growth in the likes of population and travel demand. Not only are such forecasts and associated "predict and provide" or "trend planning" initiatives open to changes in market conditions in the lifetime of plans that allocate land for development, they are often not starting from where they thought they were.[15] However, state interventions can also manifest in altogether larger discrete or systematic interventions in which different tiers of government are co-implicated, such as individual mega-infrastructure projects and national infrastructure building programs that systemically distort patterns' accessibility and development potential, with barely foreseen intergenerational consequences. In and of themselves, such large, discrete interventions have commonly come to be regarded as great planning disasters owing to their many unintended or unanticipated effects.[16] So it is with Tom Wolfe's depiction of the edge city of Perimeter Center emerging amid suburban residential Atlanta, quoted in the epigraph to this chapter.

A key feature of such planning interventions—irrespective of their scale or the tiers of government involved—is that they themselves produce unintended or unanticipated consequences. Moreover, these unintended effects of state intervention tend to become magnified over time: "The state continually recreates this contradiction [between private and collective action] at successively higher levels of complexity," write Scott and Roweis.[17] The magnification of such contradictions has a strong geographic aspect to it: space can be overcome through the production of space, but successive spatial fixes actually give capitalism's contradictions greater latitude such that, over time, the tension of overcoming space through the production of space becomes deeper and more complex.[18] Finally, state interventions produce a series of unintended consequences for the state itself. As Offe points out, "Every time the state deals with a problem in its environment it deals with a problem of itself, that is, its internal mode of operation."[19] The unintended consequences of its own actions pose a problem for the modern state in its ability to perceive problems or their nature; they pose problems for the state's continuing to "see like a state."[20]

The problem is that, while identifying the dominant role of the private sector in shaping such interventions, work in the growth machine tradition has relatively little to say about the state's subsequent unintended and contradictory impacts and their effects on the private sector and on urban politics going forward. In light of the observations above, the omission is quite significant. Related to this is the manner in which the "localist" orientation of the growth machine and urban regime concepts can obscure the intergovernmental relations and extralocal state interventions that impinge on localized growth and growth politics.[21] I describe this in table 3.1 in terms of the relationship of both the forms and tiers of government intervention to urban politics in different settlement types. The extent, timing, and legacies of state interventions significantly shape the mode of urban politics in different places, as well as in a single place over time.

The growth machine idea speaks to the prominence of private-sector interests, particularly landed business interests, in urban politics. However, even in the United States—in what I take as perhaps the best concrete approximation of a free market in land and property—state intervention has played a significant role in shaping real estate markets and, as a result, urban politics. This is especially the case with respect to postwar suburban development, which "bore the imprint of government more so than any period of suburbanization."[22] Suburbia and post-suburbia can themselves be seen as manifestations of the contradictions apparent in the urbanization of capital in which all tiers of the state are implicated to differing degrees at different times. Indeed, the state and its interventions represent a critical continuity between suburbia and post-suburbia.

Thus, to repeat a story by now well familiar to readers, America's postwar suburbs formed part of a Fordist spatial fix. That is, "post-war suburbanization has served as a vast outlet for capital in all its forms; as direct investment in factories, infrastructure and housing production; as consumer buying; as credit creation. . . . It would be very hard indeed to imagine how American capitalism would have fared if it had had to make do with cities as they were in 1949."[23] Moreover, in the numerous interests that have been brought together or incentivized for its production, suburbanization—regardless of exactly where it takes place in America—has acquired a systemic or automated, machinelike quality.[24] In important respects, then, "suburban sprawl is an idealized artificial system."[25] It is as much a product of seeing like a state as are modernist-inspired capital cities and other projects of national development, though it is rarely recognized as such. Though sprawl is often thought of as occurring "naturally" or "spontaneously," it has been thoroughly planned as a result of interventions by all tiers of government.[26]

The issue is that the suburban spatial fix has itself, in its unanticipated consequences (such as the extreme separation of land uses and the provision for automobility), become a barrier to further accumulation.[27] There is a sense here in which "it is not the crises, but . . . the victories of capitalism which produce the new social form. . . . It is . . . normal modernization and further modernization which are dissolving the contours of industrial society."[28] The emergence of post-suburbia—seen as a rounding out of traditional suburbs into cities in function but not in form—and of a new politics associated with it might be seen as embodying the search for a new (post-Fordist) spatial fix.[29] Here we see continuity between the suburbia facilitated by modern state interventions that now generate a host of unanticipated side effects as barriers to accumulation and the attempts of a post-suburban politics to ameliorate those effects through contemporary state interventions of a second modernity. Post-suburbia, or what Douglas Young and Roger Keil refer to as the in-between city, can be contrasted to urban spaces alternatively vacated by the state or produced and sustained almost entirely by the state. In this sense, post-suburbia represents a settlement form that is a "mixed product of both, state presence and state retreat."[30]

Moreover, these continuities have been paralleled by an evolution in the mode of decision making associated with a transition in state interventions from modernity to a second modernity, or what Offe terms the shift from "bureaucratic" and "purposive-rational" to "participatory" modes of decision making within capitalist states.[31] Each of these modes has its unintended side effects. Indeed, the unintended and unsettling effects of modern state interventions weaken traditions surrounding those same interventions, which in turn serves to politicize areas of life previously regarded as private.[32] This situation has led Ulrich Beck to argue that a second modernity with a "grass-roots developmental dynamic" has seen the politicization of the burgeoning side effects of modernity.[33] Arguably, this implies that participatory modes have become the signature of the state's decision-making logic in this second modernity and that politics has come to center increasingly on participatory rather than on representative democratic channels. Of course, these "participatory" modes of decision making, reflecting and responding to the politicization of such demands, will have their own side effects, if perhaps as yet only barely discernible.[34] First, as Becker and Jahn underline regarding newly emerging environmental politics, "what is at present being discussed as the ecological crisis is in essence a crisis in the societal relationships with nature and problem solution means intervention in their dynamics—with often unpredictable and unwanted, dangerous side-effects."[35] Second, the "autonomization" of the consensus-building process under such participatory modes—what planning theorists term collaborative or communicative planning—places the state under pressures that overburden it.[36] Mark Gottdiener highlighted this dilemma some time ago with respect to the evolving local politics of suburban development, arguing that "the outcome of the present pattern of political response may very well be organized politicization of every single interest in the society as the unanticipated environmental effects of growth continue to proliferate."[37]

Beck's analysis of a reflexive or second modernity is multifaceted and includes an emphasis on the politicization of the burgeoning environmental risks (the side effects) of modernity and processes of individualization in society associated with the rise of special interest groups and identity politics. What I take from Beck's analysis with a view to scrutinizing the politics of post-suburbia is the notion of the politicization of the side effects of modernity, including not least the side effects of state interventions in that period. Beck concentrates on the politicization of some of the biggest and most risky unintended consequences of modernist state interventions, such as environmental pollution and the spread of nuclear technologies. Yet the unanticipated effects of state interventions in promoting low-density suburban development can hardly be overstated, not least because of their significant contribution to inducing global climate change and their origin in the systemic properties of automobility.[38] In this respect, sprawl "wasn't an accident, but neither was it based on a specific vision of its physical form or of the life that form would generate."[39] The system of automobility expanded on the basis of what economists refer to as positive "network externalities," in which small events can lock in industrial development to a particular path.[40] It hardly needs pointing out that the same path dependencies have generated negative externalities that are equally broad in scope. These include the manner in which urban planning, architecture, and design have conformed to the needs of the car in the honing of the suburban development format, but also the subsequent contribution of extensive patterns of travel to vehicle emissions and hence climate change.[41] These negative externalities are what Lars Lerup has referred to as a "toxic ecology" of suburban sprawl "whose consequences flow downstream," a toxic ecology that the market is unable to deal with and that, in America, is in need of airing in a public sphere so that its effects can be known and possible solutions can be put forward.[42] The consequences can also hardly be overstated, in light of how residential and associated consumption patterns in the city-regions of developing countries are rapidly converging on those found in developed countries as the export of the U.S. suburban lifestyle—actually a very recent phenomenon—begins in earnest.[43]

In short, one key aspect around which the new post-suburban politics (as an emerging response to the side effects of modernist suburbanization) will coalesce has to do with what has been variously termed the urbanization of suburbia, the "retrofitting" of suburbia, and the repair of sprawl.[44] These ideas form something of a *Zeitgeist* in architectural, design, and planning circles; and to the extent that they have some substance and continue to gain traction, they have the potential to reshape the politics and ideology associated with suburbia, and with that to rework suburban into post-suburban forms.

Ellen Dunham-Jones and June Williamson suggest that "the systematic development of suburban sprawl was the big architectural project for the last fifty years . . . the redevelopment of sprawl into more urban, more connected, more sustainable places is the big project for this century."[45] This view is to an extent borne out by evidence of gradual increases in the density of suburban tranches of metropolitan spaces across the United States. "The standard wisdom about accelerating sprawl in American cities and suburbs is misleading. Within the city and suburbs, the national trend is toward a slowing if not reversing, of the long decline in density that took place earlier in the twentieth century," writes Robert Bruegmann.[46] It is also a view bolstered by academic and popular commentary on some of the demographic trends in the country.[47] Thus, Joel Kotkin in *The Next Hundred Million* has this to say: "Over the next few decades . . . suburban communities will evolve beyond the conventional 1950s style 'production suburb,' vast housing tracts constructed far from existing commercial and industrial centers. The suburbs of the 21st century will increasingly incorporate aspects of preindustrial villages. They will become compact and self-sufficient, providing office space as well as a surging home-based workforce."[48]

Underlining some of the fundamental continuities in any sequel to suburbia in America, Aaron Passell has argued that the New Urbanism movement "is dialectically related to post–World War Two suburban development: the conditions that gave rise to what has become conventional suburban development, gave rise to the conditions out of which the New

Urbanism constituted itself."[49] The New Urbanism movement has made powerful and universal counterclaims regarding the properties of good developments. It has been energetic in its attempt to convert municipalities through its form-based zoning code, with some success nationally. Moreover, the suggestion that "the locations in which these developments seem to flourish are sprawling, de-centered regions, somewhere between suburbs and exurbs, lacking clear geographic focus" is indicative of the prospects for such the New Urbanism–inspired developments to fashion a measured density and centrality and connectivity within suburban expanses.[50]

It seems certain that some of this retrofitting might well be insubstantial in nature, a consciously postmodern affectation in which "'city centers' become almost an externality of fragmented urbanism . . . grafted onto the landscape as a[n] . . . afterthought by developers and politicians concerned with identity and tradition."[51] Moreover, debate rages over whether the New Urbanism has actually fueled suburban development through the creation of so-called lifestyle centers. Perhaps more charitably, the New Urbanism is "neither a revolution nor a rejection of suburban development, but rather a pragmatic realignment."[52] In other instances, however, such retrofitting is potentially much more substantial, signaling the recasting of new settlements thought to lack sufficient agglomeration economies or to transform placeless settlements into more self-contained urban places. There is also the possibility that such experiments may prove to be profound, and exemplars of the broader possibilities for the repair of suburban sprawl.

The lack of strong political leadership, neighborhood resistance to change, and a preference for slow-growth policies are cited by Dunham-Jones and Williamson as among the causes preventing the urbanization of, for example, edge cities.[53] To these can be added the localism that pervades intergovernmental relations and hampers the suturing together of new arrangements at the sorts of county and regional scale now thought necessary to address the uneven distribution of the costs and benefits of urbanization and to facilitate the extensions of water, sewer, and public mass transit infrastructure necessary to the long-term sustainability of outer suburbs. The post-suburban politics of suburban retrofit is not simply a matter of local politics but involves as well the mobilization of political and administrative resources and imaginaries across multiple jurisdictions at larger territorial scales.

Urban Politics, Structural Change, and Settlement Evolution

To this point I have emphasized the diversity of different settlement types and the various pathways settlements can follow as they evolve. However, this is not to argue that each scenario generates a distinctively different urban politics. Rather, the autonomy of urban politics—central to both the growth machine and the urban regime concepts—is a relative one, such that we need to specify the place of urban politics within the larger process of capital accumulation.[54] However, the traditional frameworks for examining urban politics are poor at specifying these relations, reflecting both their inductive origins and the weight they attach to local agency. As Harding notes, "in their enthusiasm to underline the importance of human agency in urban development . . . they pay less attention to historical and structural factors that predispose local public officials, as well as profit-seeking business interests, to support growth strategies and the interest coalitions that promote and sustain them."[55] This in turn poses problems both for analyzing change over time and for considering differences among settlement types. The differences necessarily entail a parallel analysis of the role of the capitalist state in shaping land and property markets. The problematic for urban politics is, as Todd Swanstrom has suggested, not a question of growth versus no growth—as posited in the classic growth machine model—but rather what kind of growth.[56] Urban regime theory, as Alan Harding notes, is more sensitive to how growth politics vary over time and space, although empirical studies have typically focused on single, stable regimes.[57] Neither of the traditional concepts is linked in explicit analytical terms to a conception of structural change in capitalism or to different classes of settlement.

There are various ways in which we can conceptualize historically the processes of capital accumulation and their contradictions, including short-term business cycles, longer-term regimes of accumulation, crises, and broad secular trends.[58] The question then becomes how best to interpret the mode of urban politics in relation to these aspects of structural change so that regulation theory becomes more capable of "accounting for the dynamic spatial diversity in unity" apparent in processes of accumulation.[59] We know, for example, that the business cycle can precipitate changes in the character of local political representation, although growth machine and urban regime theory are premised on the fact that such short-term movements are unlikely to have a determinant bearing on the coalitions of interests at the center of urban politics.

In examining the relationship between urban politics and structural change, we might note how the growth machine concept surfaced during a period in which the Fordist regime of accumulation had reached its limits.[60] It is little wonder that the growth machine concept emphasizes exchange values over use values, and largely neglects issues of consumption, including state intervention for collective consumption and the reproduction of labor. Much of the post-1945 growth in advanced capitalist economies, not least the United States, was produced from the suburbanization of capital. It follows that the classic growth machine politics of the Fordist era was not strictly urban per se but rather *suburban:* the intensive Fordist regime of accumulation was underpinned by land-extensive patterns of development and accumulation in the secondary circuit of capital released by way of significant federal and state interventions licensed by suburban growth machine politics. It is also in suburban settings that landed interests have tended to be least encumbered in the development process and, moreover, are often aided and abetted by government and local politics. Accordingly, we can argue that the growth machine represents a mode of urban politics most closely associated with new suburban development in the Fordist era (see table 3.1).

There are also secular changes in the urbanization of capital that can usefully be linked to the mode of urban politics. The growth machine concept pays particular attention to the geography of development, urging us, in Harvey Molotch's words, "to see each geographical map—whether of a small group of land parcels, a whole city, a region, or a nation—not merely as a demarcation of legal, political or topographical features, but as a mosaic of competing land interests capable of strategic coalition and action."[61] However, we know that geographic patterns of urban land ownership and lease evolve over time. Settlements represent patchworks of private, club, and public realms that tend to become more complex over time, serving to make the development process itself more protracted and longer term for both the private and public sectors.[62] The growing complexity of property ownership and leasing may serve to militate against strong ideologies associated with development. It also suggests a relative shift in the collective interests of a growth coalition from the exchange values of land and property toward their use values. Greater complexity, as I indicate in table

3.1, is likely to favor modes of urban politics that resemble urban regime–style arrangements whereby "in a world of limited and dispersed authority, actors work together across institutional lines to produce a capacity to govern."[63] In this respect, power lies in the capacity to allocate small and frequently quite complex opportunities relative to those that have existed and continue to exist on the crabgrass frontier.[64]

The suburban growth machine, perhaps the purest form of Molotch's concept, seems to be the most likely mode to have contributed to future deficits in the infrastructures required for collective consumption and production. Indeed, the very fact that many residential suburban jurisdictions have, in the process of becoming suburban, sought to increase their tax base in order to pay for utilities and social services such as schools and other elements in the makeup of viable communities tends to confirm that traditional residential suburbs lacked facilities from the outset.[65]

However, suburbs evolve and mature. Accordingly, there is considerable diversity in the contemporary trajectories of different types of suburb such that it is simply not possible to speak of the suburbs as sharing a single unified interest.[66] Leaving aside the difficulty of locating and bounding some post-suburban forms, as discussed in the preceding chapter, "even if we restrict our attention to those communities which are self-governing, we quickly observe that American suburbs come in every type, shape, and size: rich and poor, industrial and residential, new and old."[67] In very general terms, and for the purposes of elaborating on the simple scheme presented in table 3.1, there is a set of (mostly inner) suburbs that have been subject to significant economic decline and a set of (primarily outer) suburbs whose economic and fiscal position is healthier.

Moreover, it may be possible to see that some of these suburbs are evolving from affluence or decline toward a class of distinctively *post-suburban* settlements engaged in a local politics of retrofit.[68] In this way, local interests are connected to wider tiers of government to make good shortfalls in expenditures on collective consumption and production while in the process attracting citylike functions and qualities. In this particular category of post-suburban settlements, some of which began life ostensibly as employment centers and have evolved toward more balanced communities or mature suburbs, the conservative politics and ideology of suburbia have been adulterated.[69] These suburbs have often become visibly more urban, beginning to grow upward at greater building densities. One word

Table 3.1
The Urbanization of Capital and Changing Modes of Urban Politics

	Settlement type					
		Mature suburbs				
	New suburbs	**Declining**	**Stable/ affluent**	**Post-suburbs**	**Cities**	**Cities experiencing significant decline**
Mode of urban politics	Suburban (pure growth machine)	Progressive or development regime	Caretaker or antigrowth regime	Development regime	Regime politics	Growth machine or development regime
State intervention	Potential value of land is affected by significant nonlocal state intervention.	Local state is unable to maintain collective consumption expenditure.	Local state is concerned to preserve the status quo.	Potential land values are altered by piecemeal local state intervention.	Local state struggles to meet collective production and consumption expenditures.	Local and nonlocal state intervenes strongly to facilitate private-sector redevelopment.
Land use	Extensive: raw land converted to low-density urban land uses.	Extensive: stable, low-density urban land use	Extensive: stable, low-density urban land use	Intensive: use of urban land intensified	Intensive: use of urban land intense (near or at “best use”)	Intensive: use of urban land intense, but with large pockets that are devalorized

Table 3.1 (continued)

The Urbanization of Capital and Changing Modes of Urban Politics

	Settlement type					
		Mature suburbs				
	New suburbs	**Declining**	**Stable/ affluent**	**Post-suburbs**	**Cities**	**Cities experiencing significant decline**
Patterns of land ownership and use	Land ownership and lease patterns are simple.	Land ownership and lease patterns are mixed.	Land ownership and lease patterns are mixed.	Land ownership and lease patterns become complicated.	Land ownership and lease patterns impede radical redevelopment.	Land ownership and lease patterns are reconfigured by the private sector and the state.
Ownership of capital	Local	Local	Local/ national	Local/ national/ international	Local/national/ international	Local/national/ international
Term over which urban capital is fixed	Short term	Medium term	Medium term	Medium term	Long term	Long term

of warning, this one from Logan and Lolotch, might be sounded at this point. "It could just be the same old growth machine but with a decorative skin. Higher density has always been a scheme for growing rents; developers consistently lobbied for more on less. They didn't give a hoot about environment or social diversity. Where densities have been low, it has been because market demand was low. The new 'smart growth' mantra may turn out to be just another smoke screen for making more money."[70]

Settlements that might be characterized as fully urban (whether city or post-suburb) are, I suggest, more likely to be characterized by regime-style politics centering on the struggle to maintain existing expenditure commitments to collective consumption and production needs, as well as by a greater interest in issues of amenity and the use values of land. For cities, then, we can also entertain some oscillation in the mode of urban politics in that growth machine and urban development regimes may be reignited during periods of intense economic crisis. This oscillation has perhaps been most pronounced in cities, but it stands to reason that it will also be felt across the suburbs and post-suburbs of metropolitan regions listed in table 3.1 precisely because of the way in which their fortunes are interrelated in the unity of the urbanization process. Indeed, the massive flow of investment to the suburbs during the postwar period but also before has been regarded as the initial root of a major devalorization of inner-city urban land, creating a "rent gap" incentive driving processes of gentrification and comprehensive redevelopment.[71] As indicated in table 3.1, local and nonlocal state expenditures are often vital to this spatial switching of capital through the creation of special-purpose development and delivery organizations, improving locational accessibility through the provision of new transportation and communications infrastructure, and the purchase, assembly, and servicing of land in the form of parcels available to developers.

The Time and Place of Post-Suburban Transportation

The organization of the land development process that has produced the suburban expanses of the United States has a number of distinct features. Here I can only touch on a few of these. Famously, one version, Frederick Jackson Turner's, of the idea of the frontier in American history continues to exert an influence today for articulating of a sense of the limitlessness of nature to be tamed and land available for development, a sense that is scarcely curtailed by a generalized weak legal and planning policy basis for urban containment.[72] This is an understanding of the frontier singular, a phenomenon that unifies the history of American settlement. It is no accident that Joel Garreau's recent depiction of the rampant land development involved in the emergence of numerous edge cities made reference to life on the new frontier.[73] A strong sense of the legal protection of private property rights has undergirded both a greater commodification of land and prioritization of the claims of individual property owners over and against the claims of government and the wider public interest compared to elsewhere.[74] The fiscal autonomy of local government provides a much more powerful incentive for local government to share an interest with private-sector developers in the land development process as a growth machine than is common elsewhere, with the possible exception today of China.[75] Add to this the specific incentives that have existed in the form of mortgage relief and the systematic distortion of accessibility at the regional scale prompted by state and federal interstate and beltway building and one has a heady cocktail of factors driving the suburban development of America. These forces, felt commonly enough across the United States in the promotion of suburban development, are underscored by simple and singular aggregate measures of urban sprawl, such as population density, that tend to give the impression of fundamental similarities in processes of urbanization. Here, then, "Southwestern cities are just about at the national average in population density, suburbanization and automobile ownership."[76] As such, "sprawl is often presumed to be the dominant form and process of American urbanization today," yet a "simple density measurement obscures the rich diversity of experience across metropolitan areas."[77]

And so Turner also wrote of the frontiers in America in the plural, such that it is also true to say that important contradictions in American urbanization have produced "a fragmented sense of what urbanism in America is."[78] As a consequence, and notwithstanding the systemic and machine-like properties of the production of traditional suburbia noted above, the significant continuities apparent in the emergence of post-suburban settlements and a distinctive post-suburban politics nevertheless will play out very unevenly in the context of different vintages of urbanization and different formal and grassroots local political and bureaucratic cultures apparent across the United States. It should not be forgotten that not only is

American urbanization unique in an international context but also, despite its relative youth, America contains enough variety that it is important not to overgeneralize from developments in one urban setting such as a Los Angeles or a Chicago to another setting. As one commentator on urban affairs notes, "When I look at Chicago and Los Angeles . . . I see not two competing models of the future of cities elsewhere in the world, but rather the uniqueness of American (i.e., U.S.) culture, values, institutions, and policies. They represent massive physical and social constructions undertaken in special places and environments, during given time periods, and under differing conditions. . . . Both cities are period pieces. . . . They are what they are. But what they are not, in themselves, are urban models for anywhere else."[79] This is true of urbanization in the modern and second modern eras and the processes of suburbanization and post-suburbanization with which I am concerned because of the way these are shaped by these distinctly different historical legacies of urbanization, including the local state interventions implicated. One contribution in a recent comparison of the implications of urbanization for urban theory therefore notes that the three city-regions of Chicago, Los Angeles, and New York "experienced their most rapid phases of economic and demographic growth at different historical points under different economic and technological conditions. This gave each a distinctive physical form reflecting that period, *which in turn shaped subsequent physical development patterns.*"[80] The largest and most rapidly growing suburban cities—dubbed "boomburbs"—have a distinct geography skewed away from the Northeast and Midwest of the United States.[81] Moreover, the modern and second modern eras have seen significant differences in metropolitan government, intergovernmental relations, and the regulation of land development emerge across the states. In a recent review of planning policies, Timothy S. Chapin noted that "the growth management efforts that arose in response to problems associated with urban sprawl exhibit great variety. This variation makes sense given the diversity of political climates, resource constraints, topographical challenges, and cultural characteristics found across the country."[82] Finally, differences are also evident in the architecture and urban form of America's different metropolitan regions, with the Sun Belt cities considered the most comprehensible and visible when compared to the older cities of the Northeast and Midwest.[83] All of these differences have major implications for the portability of particular solutions to suburban development.[84]

There are major contrasts to be drawn, for example, between the city-regions of the East Coast, with different vintages of pedestrian, railroad, and streetcar urbanization, and those of the historically less developed South and West, where most urban development has taken place in the era of the automobile, air travel, and even the partially travel-displacing information and communications technologies (ICT). It may be the metropolitan regions of the oldest vintage that present the greatest possibilities for the reworking even of new outer suburban settlement space by virtue of the more compact nature of outer suburban development and the latent public transit and other infrastructure potentials born of previous eras of urbanization. Conversely, it may be the newest outer suburbs of metropolitan spaces that prove the hardest to transform precisely because of the relative lack of existing or past fixed infrastructure development. However, the full implications of the combination of transportation and ICT technologies for the organization of urban space are, from the current vantage point, far from clear. It is unlikely they will lead to simple binary outcomes, such as the rejuvenation of cities or suburbs, but rather will contribute further to the increasing social, political, economic, and ethnic and class-based complexity of the variety of communities that make up metro regions.

So goes it, then, with those elements that might be regarded as post-suburban. It should be kept in mind that the edge cities that sometimes are taken to be synonymous with post-suburban settlement space have diverse origins and have evolved differently. Some, such as Tysons Corner, have grown steadily and concurrently with residential development as a result of the actions of private-sector growth entrepreneurs and the sorts of pragmatic local political bargains struck between residents, governments, and business interests to balance suburban ideals with fiscal realities, which can be found across the United States.[85]

Yet others have emerged after the completion of suburban development and are superimposed on an existing surrounding framework of residential development, as in the case of Kendall Downtown, discussed in chapter 5. In this latter category it should be remembered that "some suburbs were composed of the essential elements of urbanity from the start—diversity, connectedness, a public realm. . . . It seems reasonable then to develop a definitional language of urbanism that fits the suburban context that may help them evolve in a way that is more positive."[86]

Finally, a few settlements have been born post-suburban, locally conceived and comprehensively planned from the outset as a new kind of city. Such sites include Irvine, California, which has been one source for postmodern theory–inspired discussions of the city turned inside out and of California's post-suburban landscape.[87]. They might include the likes of Reston, Virginia. The Village of Schaumburg, discussed in chapter 7, is another example. Other settlements have been comprehensively planned for as part of the metropolitan-wide distribution of development and infrastructure provision, as in the case of Phoenix, Arizona.[88] These are America's privately financed and master-planned equivalents of the British New Towns, but they are also distinctly suburban in layout, density, and a provision for automobility, as well as often being conceived to play a far larger and more autonomous role within their respective metropolitan space economies than the British New Towns typically were. Some of these overgrown suburban cities have been regarded as accidental cities, though there is enough in their adaptation of small government to deal with bigger-city realities, in their accommodation of commerce to residence, and in their dealing with important issues regarding collective consumption expenditures to suggest they have been settlements distinct from the residential suburbs of the early postwar era.[89] Whether they will also be able to come to terms with the increasing latitude of the environmental and collective consumption politics associated with suburban development remains unclear.

While suburbs share a distinct low density of development, curvilinear stem-and-leaf road patterns, and morphologies that cater to the automobile, their differing origins are intriguing for what this implies about the different prospects for the reworking of suburban settlement space. Moreover, these different origins actually speak to a complex geography of post-suburban potentials across America as a result of the contributions that the vintage, scale, and infrastructure legacies of extant suburbanization may make to the prospects for any sequel to suburbia.

In certain fundamental respects, it is hardly possible to view suburban sprawl as unplanned. However, despite the laments over the "unplanned" nature of the suburban sprawl exhibited by some edge cities, in important ways they may be the most amenable of suburban settlements to a reworking of their physical space, land-use patterns, and transportation access, and in even less tangible properties such as amenity value, sense of place, and economic development potential, when set against the "edgeless city" format that now appears to predominate. Those settlements and communities born post-suburban may simply present too large an expanse of low-density development, with extreme separation of land uses, and so poorly served (or only potentially served) by public transit as to be generations away from real transformation. In these regional-scale expanses the obstacles to retrofitting and increasing the density of relatively compact edge cities, let alone the vast expanses of residential suburbs in the United States, are immense. Chris Leinburger has suggested that suburban transportation and utility infrastructures are both difficult and costly to reengineer or redevelop to support a greater density of development and population.[90] In any case, such retrofitting assumes a political will and popular support, which are questionably forthcoming in a context in which the vestiges of suburban politics and ideology remain significant.

Conclusion

The suburbs provided a spatial fix for American capitalism, to the point that the early postwar dominance of the American corporations domestically and internationally can hardly be separated from the suburbanization of American society. Yet almost immediately, many of those residential suburbs ran up against their own limits as viable settlements in their own right, prompting a subtle and gradual mutation of suburban ideology and politics. As a result, a class of post-suburban settlements might be said to have emerged alongside historical cities and suburbs of different sorts as part of the increasingly complex landscape of settlements of different socioeconomic, political, and ethnic complexions that now exists in the metropolitan regions of America. To ignore this reality is to greatly underappreciate the potential to develop a more finely crafted understanding of cities, urbanization, and the urban question. However, the existing weight of suburban politics and the emergence of post-suburban politics also cannot be ignored as increasingly significant political and policy forces in the makeup and governance of America's metropolitan regions.

4 Politics and the Private and Collective Dynamics of America's Post-Suburban Future

The financial viability of our modern suburbs was . . . flawed from the start: the lower-density pattern of development doesn't yield enough tax revenue to pay for the infrastructure needed to support them.
—Leigh Gallagher, *The End of the Suburbs*

In chapter 2 I suggested that the third of three dimensions along which to distinguish a post-suburban America, namely, the shifting ideology and politics associated with a constellation of actors involved in the production of suburbanism as a way of life, was perhaps the most satisfactory. That is not to say that history and geography are not important. While there are some limitations in defining post-suburbs in terms of a historical era, it is nevertheless the case that modern state interventions in the creation of the suburbs have historical legacies. Similarly, while there are limits to defining post-suburbs in terms of the geographic scale of their form, it is also clear that post-suburban political sensibilities revolve around dilemmas that are necessarily metropolitan or urban regional in scope. In the previous chapter, I also positioned a distinctly post-suburban politics within the evolution of the complex settlement spaces that are our largest metropolitan regions. In this chapter I pay closer attention to the range of actors and the interests involved in the process of suburbanization in the United States and hence the prospects for sustained collective responses to the contradictions of the suburban development format in the fashioning of any post-suburban future. Since suburbanization has been driven systemically by a broad constellation of actors and interests, any post-suburban future will necessarily entail the engagement of similarly broad constellations of actors and interests to resolve a series of important political dilemmas.

Here I summarize these dilemmas in terms of the contestation between political rationalities of private accumulation (growth) and environmental conservation, the contestation between the pursuit of private accumulation (growth) and provision for collective consumption, and finally the contestation over the appropriate governance arrangements—including the geographic scale of those arrangements—needed to effect a distinctly post-suburban political response to the contradictions of suburbia. As we saw in the previous chapter, the contradictions of suburbanization have deepened and gained greater latitude over time: latitude in terms of the breadth of issues facing suburban communities and interrelations among them and latitude in terms of the geographic scale at which such issues manifest and at which policy solutions and governance arrangements may need to be found. Here it becomes apparent that these dilemmas may pull actors in different directions. As Roger Keil notes, mainstream propositions such as the New Urbanism will need to be politicized to the seemingly opposing ends of social justice and ecological change.[1] Yet against this proposition, Robert Gottlieb sees precisely a combination of advocacy for urban nature and social justice in the grassroots politics that is reinventing Los Angeles.[2]

To begin, I discuss how the various actors and interests involved in the production of suburbia have also been drawn into the production of its sequel—settlements evincing a distinctly post-suburban politics and ideology. I then discuss the three main points of contestation I have identified.

The Production of Post-Suburbia

Across the two competing perspectives of continuity and discontinuity, there is a common interest among laypeople and researchers alike in the popular ideological content of suburbia and post-suburbia. The conservative ideological content of suburbia has become quite familiar in many but not all Western national contexts as the bourgeois utopia formed by the pursuit of personal and economic freedom and escape from the city. For Dolores Hayden, suburbia is "a landscape of the imagination where . . . ambitions for upward mobility and economic security, ideals about freedom and private property, and longings for social harmony and spiritual uplift" are played out.[3] These values have been seen as an immutable expression of individual preferences common to different periods of history and across different national settings.[4] However, it is probably more accurate to say that they apply especially to suburbanization in the UK and America. As a mass aspiration, the suburban way of life shows few signs of disappearing in America, though it has been adulterated in many instances such that it now coexists with a post-suburban politics and ideology that Robert Fishman sees as marking the drawing to a close of the suburban era.[5]

As Hayden also notes, then, suburbia is also "the hinge, the connection between past and future," which clearly indicates the possibility that the ideological content of and politics associated with the production of suburban landscapes may have mutated over time.[6] Fishman and Jon Teaford have been most explicit in detecting a change in the traditional meaning of suburbia. As Teaford argues, "Basic to the emerging post-suburban polity is the tension between suburban ideals and post-suburban realities. . . . Economically these areas may have become post-suburban but intellectually and emotionally they were solidly suburban."[7] Rob Kling, Spencer C. Olin, Jr., and Mark Poster see the expression of this change in the shift from the provincial nature of suburbs to the increasingly cosmopolitan complexion of post-suburbia.[8] Something of this is also captured in what Michael J. Dear and Nicholas Dahmann refer to as the ideological, cultural and political "hybridization" associated with the contribution of international migration to contemporary urbanization, much of which takes place in the form of processes of suburbanization.[9] Even Robert Bruegmann, whose arguments stress the common, long-standing causes of suburbanization processes, is careful to note the potential transformation of ideological content: "One of the oddest aspects of the anti-sprawl campaign is the way it has altered the relationship between progressive and conservative ideas. . . . The anti-sprawl movement is a powerful compound of this new progressivism and a traditional conservatism."[10]

The emphasis on the residential preferences of individuals perhaps disguises the roles of the state and property developers and a wide variety of other business interests involved in the creation and molding of such preferences over time. The activity of organized business interests alongside the state in promoting Fordist suburbanization has been highlighted in a number of accounts, though certainly with regard to major road-building schemes these business interests were more fragmented than is often appreciated. In the postwar era, mass suburbanization provided *the* predominant spatial fix for a Fordist regime of accumulation.[11] Arguably, Fordist suburbs were the location for Harvey Molotch's "growth machine" politics, led by land-based business interests but enabled by national and subnational state infrastructure investments.[12]

Beyond this one might question whether the depth and breadth of changes toward post-suburban realities have been sufficient to sustain collectively organized business interests across a range of industry sectors in settlements. Aspects of the urbanization of suburbia had been noted as early as the 1970s by Louis H. Masotti, who suggested that "while many of the older, established, and affluent suburbs are able to maintain their 'residential only' character . . . some of the older, and all of the new 'frontier' suburbs have tried to provide for industrial parks, office complexes, major retail (shopping) centers, or some combination of the three."[13] The "pastoral capitalism" of research and office parks and campuses in particular represented a suitable accommodation to suburban residential sensibilities and was itself "a major venture in suburbanity, as influential and widespread as residential suburbs," Louise Mozingo writes.[14] Thus it was that, in the words of Margaret Pugh O'Mara, "the process of high tech growth was actually a process of city building. The suburbanization of science in the late twentieth century helped to *urbanize* American suburbs making these places closer to the classic definitions of cities in terms of their economic diversity and self-sufficiency. No longer adjuncts to the central cities around which they grew up, the high-tech suburbs of the early twenty-first century are a new and influential kind of urbanism."[15] Fishman described this new urbanity in terms of the emergence of the "technoburb," arguing that "if there is a single basic principle in the structure of the technoburb, it is the renewed linkage of work and residence" and that "decentralization had so dispersed the core functions of the old central cities that many sprawling regions acquired the critical mass of population, jobs, and specialized services to function as 'new cities.'"[16] Notwithstanding that suburbs have rarely been as residentially or industrially monofunctional as is often presumed, it is this balancing of economic, residential, and other functions that became apparent as early as the 1950s in the United States and that Jon Teaford took to be the signature of post-suburbia.[17] The inherent failure of the modern American suburb to pay its way, as the epigraph to this chapter pinpoints, was apparent early on and has driven a measure of transformation among those suburbs.

The suggestion that post-suburbs represent cities in function but not in form also implies a greater role for collectively organized business interests in post-suburban politics than might be assumed. It seems clear from previous work that organized business interests *may* play a significant role both in the transformation of *some* suburbs and in the continuing evolution of *some* post-suburban settlements.[18] Indeed, some of the potential for the increased organization of business interests in the post-suburbs is registered in their integration into the national and international sphere.[19] In 1969, only 11 percent of the nation's largest companies were headquartered in the suburbs; a quarter of a century later that figure had jumped to nearly 50 percent.[20] Against this increase should be set the finding that most commercial development across U.S. cities remains a parochial affair despite the involvement of a "transnational capitalist class" or "global intelligence corps" in development projects in world cities.[21]

The problem is that, despite the recognition that the bulk of post–World War II development in advanced nations has been outside cities proper, the ownership and organizational structure of the suburban (and post-suburban) economy and the role of political coalitions in driving the development of these settlements remain little understood, including specifically the role of organized coalitions of business interests. The emerging economic dynamics of post-suburbia are fundamentally in between in character: being a product of a complex combination and delicate balance of the forces that promote concentration and dispersal, being fundamentally interurban in location.[22]

The shifting ideological and political content of suburbia and post-suburbia, along with the changeable role of business interests in suburban and post-suburban politics, is also inextricably linked with the role of the state in facilitating development; indeed, the subsequent implications of state interventions—and yet the vitally important role of the state in the production of post-suburbia—have barely begun to be explored. Writing from the perspective of an observer of the scattered, in-between form of development apparent in some parts of Germany, Thomas Sieverts argues that "the contemporary discussion is still too much restricted to the form of the city: here the compact European city, there the dissolved American urban sprawl. . . . By contrast, a comparison of the political goals and processes of town planning could be productive."[23] The role of U.S. federal government housing and infrastructure programs in promoting employment and residential decentralization has been widely invoked in the literature.[24] Seen over the longer term, local and national states' involvement in suburbanization represents both a discrete moment (the last 100 years or so) and a relatively unique distillation both of certain individually held ideological values and collective (corporatist) interests.

Moreover, there is an important thread of continuity in the pattern of state intervention in that much suburbanization (in the United States, postwar mass suburbanization provides the example) was driven by modernist state interventions, which systematically transformed the distribution of locational advantages available to developers and residents.[25] The contradictions of these interventions—usually cast in terms of concern over the sustainability of suburbanization and sprawl—have become apparent today. Past state interventions often had the paradoxical consequence of producing patterns of development that approximated to "nonplanning."[26] The contradictions of past modernist state interventions promoting suburbanization have their recursive effects in a distinctly post-suburban ideology and politics of retrofit as part of a second modernization, in which there has been a "politicization of side effects."[27] At one level, this politicization of the side effects of modernity is evident in a greater sensitivity to the practice of planning, urban design, and community building. There is, for example, interest in the possibilities of retrofitting suburbia in the United States.[28] To be sure, some of this retrofitting might be regarded as insubstantial. Regardless of the content of these developments, however, they could be thought of as an indication of renewed recognition that urban amenity drives economic growth.[29]

At the very heart of the story of any sequel to suburbia are some important political tensions, which, although not exclusive to post-suburban settlements, arguably come into sharper focus there. Three seem to be of particular significance here: the pursuit of growth versus conservation of the environment, the tension between the pursuit of private accumulation by businesses and households (growth) and the provision for collective consumption, and the contradictory pressures toward government and governance as a means of addressing the ever-increasing geographic latitude of dilemmas surrounding the environment and collective consumption.

Moreover, it is in respect to this third dilemma—the governance responses to the previous two dilemmas—that I draw a distinction between mark I and mark II post-suburban political sensibilities. Mark I post-suburban politics is constituted by what Jon Teaford identified as the early and purely local response to the need to underpin the fiscal base and hence pay for the growing burden collective of consumption expenditures of suburban communities. Here it appears that, irrespective of the precise trajectory or scale of suburban communities transitioning to post-suburbs, the limited liability of suburban politics endures in this purely local adaptation of suburban ideals to immediate economic realities. However, since the geographic latitude or scope of many of the collective consumption and environmental dilemmas commonly exceeds the local (city or county) scale, it might be suggested that a mark II post-suburban politics entails a "global sense of place," a maturing and broadening of mark I post-suburban political sensibilities, and the potential for a metropolitan or urban regional response to the contradictions of suburbanization.[30]

Growth and Provision for Collective Consumption in Post-Suburbia

The altered functional basis of post-suburbia when compared with suburbia—and in particular its increased economic gravitas—might be taken to signal one particular political tension. The accumulation process does not necessarily entail the adequate reproduction of the conditions for further accumulation.[31] This is so much the case that the urban question has been considered to revolve around provision for the collective consumption needs that underpin the reproduction of labor in particular.[32] In some parts of Europe and the global south, the central place of providing for collective needs has shored up a strong tradition of grassroots politics, and to an extent it continues to do so. In the United States it has manifested in more muted ways through representative and occasionally participatory political channels, though the case of Los Angeles, with some grassroots mobilization around making it a livable city, may yet come be representative of many other metropolitan areas of America as a result of migration and demographic trends.[33] Regardless of these differences, and even though much has changed since Manuel Castells wrote in the 1970s (including changes in the technical and organizational possibilities for the supply of amenities and services for collective consumption), it could be argued that this urban question was never strictly an urban question at all but most often a *suburban* question.

Parasitic Suburbanization and Post-Suburban Politics Mark I

The dilemma of balancing the pursuit of economic growth and providing for collective consumption, as Teaford documents, was apparent very early within many suburban tract communities. In this sense, and in terms of this particular political dilemma, the production of suburbia almost immediately ushered in a local post-suburban political response by attracting a business base in conjunction with incorporation or soon thereafter. Post-suburban sensibilities have been apparent from this time to the more recent and very rapid growth of very large "boomburbs" (incorporated suburban cities of more than 100,000 residents), found predominantly in the U.S. West. These major expanses of suburban density development are not simply bedroom communities but house significant concentrations of jobs of all sorts, and partly as a result of the need to balance growth with providing for local collective consumption they have been "inventive places that devise numerous strategies to adapt governments intended for small towns to the realities of big cities."[34] In this respect, write Robert E. Lang and Judith K. LeFurgy, "boomburbs offer a surprising mix of governance. Residents seek both more and less government than residents of suburbs around cities in the Northeast and Midwest. Boomburb municipal governments are usually smaller than those of comparatively sized old-style big cities, but filling the gap are often private governments, such as home owners associations and various shadow governments, such as special improvement districts."[35]

This suburban contradiction that arises out of the need to effect an appropriate balance between the pursuit of growth by attracting private residents and developing businesses and the corresponding required expenditure on the physical and social infrastructure needed highlights more than anything the way in which "economic development and collective consumption are arguably internally related but not necessarily at the local level; and the conflicts within and between those with interests in economic development and collective consumption respectively do not necessarily generate jurisdictional projects. Whether they do is a contingent matter."[36] This is not to say that such post-suburban political accommodations were not resisted. They were, although the opposition was framed in terms of a distinct though not unrelated contradiction between growth and environment, which I consider below.

It is in suburban settlements, and more particularly in primarily residential suburbs, that such undersupply for collective consumption needs has been most acute, partly because of the residential development process itself, the absence of strong government on unincorporated lands or the weakness of county governments, and many residential builders' preference for tract building rather than comprehensive community building. However, where governments exist, the undersupply of local collective consumption needs may also reflect the local government itself seeking private investment in the same manner as the private sector, which local governments sometimes take to court. As J. Persky and W. Wiewel have documented regarding the costs and benefits of employment decentralization in the United States, "Why do so many suburban municipalities aggressively seek manufacturing development if the net fiscal impacts of such development are negligible? . . . Like the businesses that focus only on their private profit, these municipalities ignore the broader consequences of their actions."[37]

Robert A. Beauregard characterizes the mass suburbanization that occurred in the postwar United States until the 1970s as a "parasitic" urbanization that was nevertheless an anomaly in terms of the longer history of urbanization before and since.[38] Interestingly, this "parasitic" mass suburbanization was still attended by critical commentary, most notably centered on the pursuit of such unencumbered growth and the associated underprovision for collective consumption. "Large postwar developments were planned and built, not as neighborhoods, but as large aggregations of a single house," with the consequent undersupply of collective consumption of all sorts.[39] "Developers were putting up houses faster than local governments could respond. Consequently, there were too few classrooms, the roads were congested, parks and playgrounds lagged behind need, and taxes were on the rise."[40] It is the need to redress such underprovision of infrastructure, services, and amenities that creates a distinctive political tension as the governments of suburban settlements seek for their communities to become post-suburban. That is, in a picture that remains true today, provision for local collective consumption—schools, fire and police services, transportation infrastructure, and so forth—was rarely concurrent with residential and commercial and industrial development for private use.

Regional Transportation Infrastructure and Post-Suburban Politics Mark II

While such dilemmas regarding the relationship between growth and collective consumption may signal something of a divide between business-led growth agendas and public-sector-led antigrowth agendas, they have also at times proved important points of consensus for "growth coalitions" at the city-region scale.[41] Settlements experiencing strong growth are routinely associated with concerns regarding "overheating" and perceived needs to make good on "infrastructure deficits," which can spill over into city region–wide politics. These are not the only collective consumption needs that commonly manifest beyond the scope of individual established city or suburban jurisdictions but they are among the most important, not least because of the unintended consequences of the major road building of the past and the weight of expectations placed on mass transit to effect a reworking of suburbia's contradictions.

The infrastructural framework for suburbanization was enhanced significantly in the early postwar years by federal government spending on the development of the Interstate Highway System and on associated state funding for beltways and parkways. Although this program of road building was quite short-lived, lasting only around fifteen years, it was intense, and fundamentally distorted patterns of accessibility not only at the metropolitan scale but also at the regional scale and to some extent at the national scale. Such massively increased accessibility in the outer reaches of metropolitan areas increased development potentials and economic growth based on the one-off differentials achievable in the exchange value of land as it was for the first time transformed from rural uses. That is, development opportunities in the peripheries of metropolitan regions were increased exponentially, but by the same token, the costs of integrating such new, ostensibly residential communities into the extant metropolitan space and governance arrangements by way of the provision of fixed infrastructure and other collectively consumed services also grew exponentially. For a time, during the 1960s and early 1970s, "postwar suburban politicians enjoyed and expected access to federal officials in matters such as paying for highway building, water and sewer development, and other infrastructure improvements."[42] Although, across a range of collective consumption expenditures such as for school, fire and police services, and water and waste collection and disposal, post-suburban realties had become apparent almost immediately in newly built suburbs, they were greatly magnified with the arrival of what M. H. Rose and R. A. Mohl describe as the "post-interstate" era, from 1973 on. By this time the unanticipated effects on central cities had become apparent for all to see, although it could hardly be said that they were in all instances unanticipated.[43]

The effects of this burst of expressway building were felt differently in both cities and suburbs. The side effects of such massive modernist engineering-led and planning-supported interventions were felt almost immediately in the centers of historical cities but came into view only later in the outer suburban expanses of the metropolitan areas of America. Arguably these side effects of such a massive distortion of regional patterns of accessibility and the development it set in train remain less visible to residents, politicians, and policymakers both in terms of their immediate implications for urban amenity and place shaping and in terms of their longer-term costs and benefits.[44] Thus, although the post-interstate era has ushered in greater debate on the virtues of, and funding for, a greater mix of travel modes, there is little momentum for additional investment in mass transit in cities and inner suburbs, and less for the outer suburbs. The considerable private economic benefits that accrue to developers and to some extent individual business and residential occupants continue to generate an associated politics of mobility centered on increases in automobility and road building in the form of outer metropolitan beltway proposals, often resurrected from those shelved at the height of the interstate era. A case in point here is the Washington, D.C., metropolitan area, which I report on later in discussing Tysons Corner in chapter 6. Thus, as Rose and Mohl write, "the constituencies that value automobility—most state and city officials, political and business interests, highway engineers, and suburbanites in sprawling metroburbs—generally accepted the need to replace aging elevated expressways, but they also wanted to rebuild, relocate, deck over, or tunnel new and even bigger expressways in order to serve growing metro areas and meet projected traffic demands. . . . In the largest sense, however, the teardown movement has been only minimally successful in changing attitudes toward behavior, or transportation policy in American cities."[45]

In sum, this contradiction between the support of private accumulation, on the one hand, and reproduction of the conditions for such growth through providing for collective consumption on the other pervades state intervention in general and planning in particular in capitalist economic systems. This political dynamic is also underlined by the ideology of post-suburbia as an amalgam of traditional suburban ideals of personal freedom and residential preference, balanced with recognition of economic and fiscal realities. This tension manifests itself at multiple geographic scales across city regions and even within the fastest-growing and wealthiest of post-suburban settlements owing to the limited purview of the purely local mark I post-suburban politics involved. It is most visible at the scale of major city regions as a result of the highly uneven distribution of private and social costs and benefits across major city regions, and it is closely related to dilemmas over the future metropolitan governance that I discuss later in the chapter.

Growth and the Environment in Post-Suburbia

One powerful contradiction of the suburban spatial fix relates to how a distinctly post-suburban politics might be fashioned among the inhabitants of traditional residential suburbia since "we are confronted with a paradox: one of the characteristics of . . . second modernity . . . is precisely the need for more participatory governance. At the same time, however, the spatial dilution of citizens in suburbia and the new urban lifestyles linked to urban sprawl may jeopardize the success of these participative endeavors, since the low density form is deprived of public spaces and stimulates privacy to the detriment of social capital."[46] There are at least two aspects to whether such a contradiction may be overcome with the emergence of what could be regarded as a post-suburban politics. The first is whether the privatism of residential suburbia precludes a pragmatic recognition of collective consumption needs at the level of individual communities. For all but the smallest, most affluent suburban communities, there is evidence that it does. The second is whether and to what extent any "global sense of place" will drive a recognition of environmental impacts that necessarily transcend individual communities and in doing so help avert a tragedy of the environmental commons.[47]

The Residential Suburb as a Barrier to the Emergence of a Post-Suburban Politics

For Joel Kotkin, "a nation that must accommodate one hundred million more residents cannot do so with a slow growth strategy. Rather, innovative technologies and new ways of organizing work may provide the key to achieving both economic growth and environmental sustainability."[48] That is, the opposition between economic growth and environmental conservation is something of a false one. Yet the reality has been that pro-growth and antigrowth or conservation agendas are crucially in tension within suburban politics. Indeed, the jealous protection of residential and environmental amenity in suburban communities is something that David Harvey highlights as both contributing to the suburban spatial fix and erecting a barrier to future possibilities for accumulation. "The American suburbs, formed as an economic and social response to problems internal to capitalist accumulation, now forms an entrenched barrier to social and economic change. The political power of the suburbs is used conservatively to defend the lifestyle and the privileges and to exclude unwanted growth," he writes.[49] This is a disjuncture between an enviable private realm and an unenviable public realm in suburbia, a disjuncture that has resulted in suburban NIMBYism.[50] In the United States, while the ideal of proximity to nature and a sense of personal freedom attracted suburban residents en masse, municipal governments' enactment of this collective sense of independence saw the use of planning powers to uphold property values and promote economic development through attracting wealthy residents rather than by preventing environmental degradation per se.[51]

In this respect, there are real limits to any emergent post-suburban politics to shape the suburban landscape since "the zoning codes and land use practices that produced the conventional suburban form of the twentieth century are simply too entrenched and pervasive for piecemeal, incremental projects to adequately improve the sustainable performance of suburbia as a whole."[52] The inherent limits to some suburbs becoming post-suburban are registered in the relatively homogeneous residential populations they house. Thus, while it could be argued that "for their long-term fiscal, economic, and social vitality, suburbs will need to become more diverse in their housing and by implication demographic options," it seems likely that the conservativism of existing suburban populations will resist such diversification.[53] For all their notoriety in professional circles, the principles of smart growth, TOD, and New Urbanism may have little purchase on the suburban landscape. As Anthony Flint rather gloomily observes, "the experience of the smart growth movement over the past fifteen years does seem to suggest one maxim: it's doubtful the citizenry will ever rise up against sprawl as a political matter, or out of concern for the environment, or from any sense of obligation to the greater good."[54] These same constraints on the emergence and efficacy of a distinctly post-suburban politics are also operating outside the traditional exclusionary residential suburbs. "'Edge cities' auto-dependent infrastructure, their relative lack of strong political leadership, the rise of neighborhood resistance to change and slow-growth zoning effects, and the demographically homogenous population that combusts with its urban core counterpart are the reasons most often cited for low expectations of edge city urbanization."[55]

Moreover, this privately oriented environmental rationality associated with suburbia contains contradictions of its own, insofar as the private and collective aspects of environmental concerns are often unconsciously or consciously elided by educated, middle-class suburbanites. Where pro-growth interests in post-suburbia are operating in a context of historical settlement patterns, it is noticeable that they have been overlain by or met with considerable opposition from conservation interests. These have become increasingly apparent as aspects of a growing collective environmental consciousness and politics emerged by the 1960s and early 1970s with concrete, although often temporary or diluted, effects on suburban and metropolitan governments. The elements of the regional-scale or state-level planning that was enacted in the 1960s and 1970s amounted to a quiet revolution that curbed some of the excesses of earlier development. Yet environmental and conservation interests have often generated paradoxical interventions in the development process.

Again, the paradigmatic instances of Southern California's post-suburbia provide further evidence of an emergent tension between the traditional meaning of suburbia and the new economic and fiscal realities driving the development of many new settlements. As several commentators have noted, California's post-suburbia has—from the outset—been planned. From its inception, such broader planning has been riddled by the interests of business.[56] Moreover, as local environmental and conservation lobbies have grown in these new settlements, it is often the case that their co-option into a highly legal-rational planning and land development system has produced perverse effects. Richard Hogan in *The Failure of Planning* argues that the urban expansion of San Diego provides a better illustration of the failure of planning and environmental concerns when set against pro-growth lobbies such as exist in Los Angeles since San

Diego is commonly regarded as a planned city. In the case of San Diego, the very strength of regional—or what Hogan refers to as "big picture"—planning operating within a legal-rational process itself became the force driving post-suburban sprawl, since it was private sector developers and corporate environmental groups that provided the expertise to plan effectively at this scale. Where grassroots opposition to continued growth had an effect, it was to exacerbate the planning problem, forcing the city government into the arms of developers equipped to provide a big-picture planning solution.[57]

A Global Post-Suburban Sense of Place?

The elision of the conservation of private residential amenity and environmental conservation captures perfectly the way in which environmental issues have tended to be located in a politics of living space or of reproduction that in turn is separated from the politics of production—a separation that is increasingly untenable if any post-suburban politics, urban form, or regional governance is to emerge.[58] For a start, the sorts of social and demographic diversity that Arthur C. Nelson sees as vital to the fiscal and social health of suburbs are no less important for what they entail regarding the future environmental quality and impact of suburbs and their reproduction. Yet suburban constituents continue to have a narrow geographic purview of how the contradictions of suburbia might be addressed in many metropolitan settings. In part, this is a legacy of the thinness of the tissue of government within which suburban development has taken place. In the early history of postwar parasitical suburbanization the interests and actions of public officials and private individuals appeared largely synonymous, to the point that local political reliance on the various actors in the land development and property markets led to an inability to conceptualize specifically the geographic scale of the implications of such suburban development. Thus, as Mark Gottdiener writes, "the parties involved with the planning process, with the exception of the professional planners, take a very limited view of regional needs."[59]

The locally rooted concerns described above as barriers to the emergence of a distinctly post-suburban politics have also been overlain to an extent by a growing suburban middle-class awareness of the broader impacts of climate change and environmental destruction. This typically is an environmental politics that is broader than the residential and environmental amenity sought by consumer-voters in their retreat to the suburbs and jealously guarded by competition among suburban jurisdictions in the classic model outlined by Charles Mills Tiebout.[60] Whether such broader concern over the environmental side effects of suburbia resolves itself into a politics that calls for collective responses that transcend the pervasive localism apparent in America is an open question. Yet it is precisely those environmental issues that transcend jurisdictional boundaries that some see as driving a new scalar politics and new inter-suburban governance arrangements. As Andrew Jonas and David Gibbs have it, a gathering awareness of environmental sustainability among local politicians, bureaucrats, and the general public has led to "experimentation with new spaces of governance that blur the distinction between 'city' and its 'region' or 'hinterland.'"[61] Here, environmental concerns related to water and waste disposal, energy consumption, and integrated mass transit may see some resurrection of the regional-scale tradition of planning in America. This is certainly the scale emphasized by some commentators as the most likely and appropriate scale at which such political and policy conversations should be had.

Beyond these questions of environmental spillovers at the metropolitan or regional-urban scale, the broadest of environmental concerns regarding global climate change and the like have, since at least the 1960s, prompted the formation and growth of local, national, and international environmental interest groups and informed local antigrowth suburban politics and the adoption of specific planning and other policies. Some evidence of the latter effects of an environmentally informed local politics and associated urban management and planning is reported in the instances of Kendall Downtown and Tysons Corner covered here, though in both cases the depth and longevity of such ideas can be questioned. However, the charge is that environmental interest groups have been co-opted into government priorities for the ecological modernization of urban sprawl that have reflected long-standing elite, notably land development, economic interests. Such priorities for the ecological modernization of urban sprawl have been narrowly conceived so that environmental interests have inadvertently failed to fundamentally question the production of urban sprawl itself.[62] Both "sustainable mobility" based on low-carbon fuels and new technology and a "sustainable urbanism" based on reworked land-use and transportation planning of cities and suburbs would be needed to effect a significant improvement in the urban transportation sector's environmental footprint.[63] Whether an emergent and seemingly altogether broader environmentally informed wave of growth management policies in the United States will pose fundamental questions of, or present constraints on, the nature of the land development process or else be a breeding ground for a stronger and broader-based ecological modernization of suburbanization, a sequel to suburbia, is as yet unclear.[64]

Again, however, there is some evidence to suggest that the same social and demographic trends that Nelson sees suburban communities having to address in the future may have their effects in a progressive post-suburban politics in some metropolitan areas of America. In particular, as Gottlieb illustrates with reference to Los Angeles, transnational migration to many metropolitan areas has not only been a factor in the increased population density of some outer suburbs, it also offers the prospect of suburban communities rooted in particular ethnic populations seeking to address local environmental and social justice causes with reference simultaneously to some of their international causes and consequences.[65] While this is likely to be a politics that is selective and piecemeal in terms of the issues addressed, it is nevertheless one that is informed with a global sense of place, and it may come to represent the sort of participatory grassroots politics that Ulrich Beck has taken to be the signature of the politics of second modernity.

In sum, the very fact that there has been an elision of what is being conserved, of environment and residential amenity with land and property values, highlights some of the inherent ideological contradictions of suburbia. Many of these ideological contradictions remain in what I term mark I post-suburban politics. Whether this in turn will lead to a mark II post-suburban politics is less clear. This is a tension that is also intertwined with that regarding the pursuit of growth and provision for collective consumption since, as Gottlieb notes, the greening of Los Angeles in the preservation or colonization of open spaces has at times rubbed up against the need to use such scarce spaces for vital public services.[66] In some respects this is the tension that most clearly highlights the new meaning and ideological content of any American sequel to suburbia, since the classical ideals of suburban lifestyle—such as the desire for space and a pastoral refuge—have mutated with the new economic, social, ethnic, and political complexion of these settlements.

The Scalar Politics of Governance in Post-Suburbia

The prospects for the reworking of suburban space are dependent on finding suitable *patterns of governance,* which may or may not entail *government.* They depend not only on the local self-sufficiency of individual communities but also on adequate intergovernmental arrangements in which local communities and their political leaders find something of their better nature within a broader pattern of governance at the metropolitan or regional scale. The difficulty of locating and bounding post-suburban settlements—owing to their sometime lack of status as separate administrative jurisdictions, their "underbounding" in other instances, or the sheer dynamism and shadow effects on neighboring jurisdictions—appears to be central to a politics centered on questions of appropriate governmental boundaries and the reach of governance or intergovernmental relations.[67] Here, then, post-suburban politics appears as a hesitant response to the contradictions of suburbanization, as Mohl explains: "The rapid post-war growth of suburbia . . . resulted in newer forms of urban political conflict in which city and suburb struggled for dominance and control. Many of the political issues of the time were spatial, or territorial. . . . Some urban regions resolved these conflicts with experiments in new governmental structures. . . . But most metropolitan areas simply muddled along without ever fully resolving the city-suburban political controversies."[68]

The Government of Mark I Post-Suburbs

Jon Teaford highlights the political independence of American suburbs as their defining feature. "Because of the strong tradition of local self-rule in the United States . . . [the] political distinction between suburbs and central city has been vital to discussions of suburban development, lifestyle and policy. American suburbs are not simply peripheral areas with larger lawns and more trees than districts nearer to the historic hub. They are governmentally independent political units that can employ the powers of the state to distinguish themselves from the city."[69] As such, the incorporation of suburbs as separate communities has often fashioned both an exclusionary aspect to local politics in residential suburbia and extreme fragmentation in the government of metropolitan space, with ensuing competition between suburban (village and city) jurisdictions and historical cities, between suburban jurisdictions and counties, and among individual suburban jurisdictions. Both these aspects of the political independence of suburbs are likely to remain powerful obstructions to the emergence of a mark II post-suburban political and policy sensibility that would shape metropolitan region governance. Thus, Teaford lends weight to the notion of enduring parasitic tendencies within a post-suburban America when noting that "the government of the post-suburban future [is] in large measure a product of the ideology inherited from the suburban past."[70] Yet he also notes that as early as the 1950s, the residents and politicians of suburbia had come to accept that their settlements needed to be more than just a refuge from cities. Suburbia had been, and post-suburbia continues to be, constructed within a thin institutional setting, with communities being incorporated and acquiring formal government structures after their initial development, straddling existing government jurisdictions, and eventually being woven into a more complex set of intergovernmental relations at the urban-regional scale. Here, then, "suburbia may have seemed an irrational crazy quilt of defensive municipal fragments jealous of their authority. But the increasingly durable threads of county coordination were holding these fragments together, and gradually as the post-suburban polity emerged, the pattern of authority would appear less crazy and more attractive to perceptive eyes."[71] What this discussion makes clear is that one important element of the politics of post-suburbia has revolved around place making within larger urban regions, raising perennial questions as to the appropriate scale of government and networks of intergovernmental relations.

Of course, novelty is one ingredient that has typically been used to define suburbs and suburban sprawl.[72] This is especially the case with edge cities that have existed on unincorporated county land, such that their successful redevelopment may founder on the lack of a government entity dedicated to financing and enforcing planning aspirations. This is partly why Ellen Dunham-Jones and June Williamson argue that "of all the types of retrofits we have studied, the prospect of retrofitting edge cities is perhaps the most daunting."[73] However, they also go on to note, "While most edge cities began in unincorporated municipalities, growing problems with traffic and inadequate infrastructure and services have necessitated the formation of various governmental and civic authorities—from transportation management authorities to business improvement districts and public-private partnerships."[74] One cannot discount, therefore, the potential for such employment decentralization and attempts to better plan or remediate particular developments to prompt the incorporation of ostensibly post-suburban communities with their own effects on surrounding swaths of established incorporated or unincorporated residential suburbs. The question is whether local self-government implies anything more than a purely local mark I post-suburban political response to the contradictions of suburbia.

The Metropolitics of Mark II Post-Suburban Sensibilities

Although the popular and political sensibilities associated with residential suburbs were presented as a barrier to the emergence of a post-suburban America, we would do well to remember it was not always thus. What Beauregard calls parasitical suburbanization had occurred well before the postwar era; it was just less visible, owing to the widespread tendency for these developments to be folded into existing city jurisdictions and service areas. Thus, "in the absence of decent sewage, water, and educational systems, land speculators looked to annexation as a sort of guarantee to potential buyers that the suburbs would eventually possess the comforts of the city."[75] Indeed, "if annexation . . . or consolidation . . . had not taken place, there would now be no great cities in the United States in the political sense of the term."[76] However, this dynamic changed as a suburban consciousness took hold and suburban services improved, and as urban and suburban developments alike took on stronger class and ethnic peculiarities and divisions.[77] Thus, if postwar parasitical mass suburbanization in the United States was short-lived, to be replaced by a return to distributive urbanization, as Beauregard suggests, it is also clear that parasitical elements persist in tensions over the appropriate scale of government and intergovernmental arrangements. As Beauregard argues, "any renewed urbanity would have to reappear at the regional scale and encompass the older, central cities, inner suburbs, edge cities, and peripheral developments of all sorts."[78] In light of what I have said about the unintended consequences of modern state interventions, this does not mean that any such regional-scale political and planning solutions would be without further unintended consequences. Indeed, there is a distinct danger that such unintended consequences may be altogether larger and broader in scope. Nevertheless, against this, localism has been a force in U.S. suburbanization since colonial times. It conditioned the development of the postwar suburbs, and, as "suburban areas struggle to overcome unanticipated . . . consequences of growth, localism fundamentally shapes suburban response."[79]

The tension between the pursuit of private accumulation and providing for collective consumption in individual urban and suburban jurisdictions, discussed earlier, continues to be magnified as population and economic activity continue to decentralize across America. Here, any new scalar politics of post-suburbia centers not only on emerging environmental concerns but also on the regionally unequal distribution of economic costs and benefits. "The decentralization of employment further and further into the outer suburbs imposes social costs of about the same magnitude as the private benefits it generates. These benefits and costs are distributed with huge inequities. The decay that increasingly characterizes central city neighborhoods and inner suburbs has long-run consequences that themselves threaten the viability of entire metropolitan areas."[80] Persky and Wiewel have highlighted the geographically uneven distribution of costs and benefits from employment decentralization in America, noting how "when a suburb snags a new plant, it gains the substantial nonresidential surplus generated by that plant. But the fiscal burdens created by associated residential development are spread widely across nearby and distant communities. The lead suburb gains dramatically even if all suburbs as a whole gain little."[81] The problem remains that there are major difficulties in instituting new—ostensibly regional-scale—governance arrangements, since such arrangements often remain partial in their functional coverage and difficult to construct politically.

The diverging fiscal capacities of local governments in city-regions has driven a quite variable landscape of intermunicipal competition, but also a renewed sense of the urgency of intergovernmental cooperation in the United States.[82] Amer Althubaity and Andy Jonas have highlighted how suburban and post-suburban pro-growth politics in California rest on a degree of cooperation with other municipalities, and in particular on other tiers of government. "What we . . . refer to as 'suburban entrepreneurialism' would arguably be extremely limited were it not for the ability of local government in suburban areas to harness state redevelopment powers . . . to lever inward investment."[83] The paradigmatic case of Southern California's post-suburbia seems to represent a prime example of the manner in which processes of interauthority competition and cooperation revolve around a central tension between the pursuit of private accumulation, on the one hand, and provision for collective consumption expenditures on the other.[84] Yet, as Juliet Musso notes, the "polycentric federalism" of Southern California defies conventional wisdom regarding geographically fixed patterns of fiscal capacity from inner to outer municipalities.[85]

One of the intriguing and barely investigated endogenous potentials for change within suburbs in America, as elsewhere, revolves around the extent to which suburbs will be breeding grounds for city, metropolitan, and national political and bureaucratic leadership. Some time ago Mark Gottdiener noted the generally limited ambitions and low visibility of suburban politicians in state and national politics.[86] The fact that, as suburbs continue to evolve, "redevelopment regimes" have nevertheless been underresearched suggests that the ambition and visibility of suburban political leaders may also have evolved.[87] More recently, then, Roger Keil has been prominent in exploring questions regarding the redefinition of city-region governance in both North America and Europe. Drawing specifically on the cases of Los Angeles and Toronto, he has argued that "the suburbanization of urban politics—whether in the form of urban secessionism or regional consolidation—has created a new political platform from which powerful political and economic actors . . . operate region-wide."[88]

In sum, at an abstract theoretical level, the state is deeply implicated in the precise scale and nature of the spatial fix inherent in the urbanization of capital—not least in how the boundaries of various tiers of subnational states come to be reworked and in how new intergovernmental working arrangements are incentivized.[89] While governmental fragmentation has been the norm across America, it is also the case that glimpses of any post-suburban future are also likely to be shaped by different trajectories of stasis, secession, and annexation/consolidation that exist among governments in U.S. metropolitan areas. While the latter processes have not been much in evidence in the last one hundred years or so, they were historically, and there is nothing to say that—as with many things pertaining to the urban—things will not come full circle. Post-suburban realities have promoted a degree of political and administrative pragmatism in intergovernmental relations that cuts across the predominant trend of localism and fragmentation that has existed in America. Here, to an extent, America continues to present a vast laboratory regarding such intergovernmental arrangements, not least with regard to the local management of urban growth, in light of the variety of metropolitan governmental arrangements, state-level planning, and environmental regulatory solutions such as urban growth boundaries, concurrency, and the like that have waxed and waned across several eras of growth management.[90]

Conclusion

The contradictions of suburbia crystallized, such that they now present a barrier to the future accumulation of capital. This barrier is acutely manifested in dilemmas over the pursuit of private accumulation (growth) versus environmental conservation, and over growth versus providing for collective consumption needs. These two dilemmas in turn underpin a third, scalar aspect to the politics of post-suburbia. In this chapter I discussed each dilemma in turn, though it is important to remember that they are fundamentally interrelated. In the chapters that follow I therefore discuss them in a variable order according to which one of the three dilemmas effectively sets the tone for a discussion of the other two. Thus, in the Kendall Downtown case, discussed in chapter 5, I begin with the environmental consequences of growth pursued through suburban development, since this has been displaced to an extent into the politics surrounding private versus collective needs in the activities of an existing metropolitan governmental machinery. In the case of the present redevelopment of Tysons Corner, discussed in chapter 6, it is the collective consumption needs of established business and residential communities that have provided the impetus for attempts to reshape the suburban environment, with implications for how such physical transformation should be governed. Finally, in the case of Schaumburg, discussed in chapter 7, it is developments in perhaps America's most fragmented and divided metropolitan region that continue to limit the horizons of suburban and post-suburban communities when seeking to address important inequalities and imbalances in revenues and expenditures, let alone the environmental consequences of suburbanization.

Across the United States, the need to address dilemmas surrounding private accumulation versus provision for collective consumption has been apparent for some time now in suburban communities that have already become post-suburban or in communities born post-suburban in what could be termed mark I post-suburban politics. The picture of mark I post-suburban politics does, however, remain quite varied. For instance, despite the significant build-up of business interests of all sorts in the suburbs, the collective organization of business interests in the suburbs remains weak. One exception here might be Tysons Corner. In this community, the role of the business voice in calling forth and materially underpinning a major collective consumption expenditure, the extension of mass transit, seems somewhat exceptional. Yet if America is to have its sequel to suburbia, organized business will surely have to play a similar role in countless other instances. Moreover, it is less clear whether concerns over the environmental and collective consumption–related contradictions of suburbia have to date driven new supralocal governance arrangements of any depth and durability (what I term mark II post-suburban political sensibilities). This is certainly the case with regard to the three cases I present in the chapters that follow. Despite the lack of metropolitan governance, new local governance institutions have been created to oversee suburban redevelopment in Tysons Corner. Kendall Downtown and Schaumburg have existed in differing and evolving contexts of metropolitan governance. Yet in each instance, those post-suburban political sensibilities that have emerged have struggled to exceed the local sensibilities. And in this sense—of the emergence of new metropolitan or urban regional, post-suburban political sensibilities—the overwhelming message is suburban business as usual.

5 Kendall Downtown: The Past of New Urbanism?

What is Florida anyway? An immense sunny toilet where millions of tourists flush their money and save the moment on Kodak film. The recipe for redemption is simple: scare away the tourists and pretty soon you scare off the developers. No more developers, no more bankers. No more bankers, no more lawyers.

—Carl Hiaasen, *Tourist Season*

This rant of Carl Hiaasen's ecowarrior protagonist contains pretty much most of the ingredients that have seen Florida, like Los Angeles and California a little earlier, come to be regarded as a paradise lost. "Curiously, for a state where visionaries often invoked the metaphor of Florida as a land of dreams, the evidence suggests that most developers and dream makers sought to transform the land into something else."[1] While this constant remaking of Florida opens up the prospect of a viable sequel to suburbia being set in cosmopolitan Miami, as yet the search for a next Miami seems altogether more forlorn than the prospect of a next Los Angeles.[2]

Of the three glimpses of America's post-suburban future presented in this book, the pattern of urban development and the management of it in the state and Miami-Dade County that figure here perhaps most clearly illustrate the sorts of unintended environmental consequences of modern state interventions stressed in chapter 3. Whereas much of Florida was once seen as a wasteland to be drained and filled before it could have value—an agricultural or a development value—now it is recognized as an extremely sensitive ecosystem that that has been greatly damaged by those "improvements" of modernity.[3] Thus, as David R. Colburn and Lance deHaven-Smith write, "Ironically, some of the attitudes and policies developed in the twentieth century to help modernize Florida now threaten its environment, economy, and quality of life. To stimulate population growth and economic development, state and local governments in Florida helped drain land for farming and construction, limited state authority to tax, imported low-wage agricultural labor from other nations, sold public lands to pay for public works . . . and offered Florida on the cheap to anyone who wanted to live there. Reversing these policies commenced in the 1970s with the adoption of a corporate income tax, growth management laws, and water permitting, and continues today with the effort to restore the Everglades."[4] Even as South Florida has been promoted as offering "places to which Americans can escape to a more exotic reality," it has also increasingly begun to be presented in popular fiction as a planning disaster, such as in Carl Hiaasen's *Tourist Season*.[5]

The draining and opening up of South Florida to development has produced what Robert E. Lang describes as "simply the most centerless large region in the nation—Los Angeles minus the focal points."[6] This landscape of "sprawl plus" is partly a function of its vintage. Jan Nijman, for example, elaborates how "Los Angeles was the logical end product of the U.S. cultural formation moving west. Miami . . . is of a completely different making, only understandable in a context of the international region, the hemisphere, and globalization."[7] The value in studying Kendall Downtown, located in metro Miami-Dade County, then, is simply that, in Gary Mormino's words, "the Florida of today is the America of tomorrow."[8]

If Miami and its suburban surrounds are, as Nijman goes on to say, both very new and very cosmopolitan in their ethnic complexion, then this is a metropolitan area whose cosmopolitan and highly transient character presents an altogether darker and unruly picture of cosmopolis than that celebrated in some planning theory.[9] The Kendall Downtown case presents a picture of the real difficulties of fashioning any post-suburban future in America. Here I recount the story of a past attempt to fashion a post-suburban future from suburbia. The Kendall Downtown area is an edge city of sorts fashioned around elements of the land-use and transportation planning of Miami-Dade County and the initiative of some segments of the private sector to engage civil society in the planning of a focal point for the sprawling, low-density residential suburb of Kendall. It has been documented as something of a success story of the retrofitting of suburbia.[10] This new downtown was inspired greatly by nationally significant New Urbanist ideas of two local practicing architects, Andres Duany and Elizabeth Plater-Zyberk. In this chapter I suggest that it represents a partial success of residential and commercial density and transit-oriented development (TOD) in a repetitive low-density sea of almost totally automobile-oriented suburban residential and commercial sprawl. The history of the downtown speaks to some of the very real political, governmental, and private corporate limits to the aspirations of what remains a minority of architects, planners, and politicians in America today. More particularly, of the three case studies presented in this book, the Kendall Downtown story also provides the starkest illustration of the political tension between growth and the environment, highlighted in the previous chapter as the tension most likely to characterize America's post-suburban future.

Sprawl in the Sunshine State

Historically, most of Florida's population was in North Florida, to the point that South Florida was virtually uninhabited. In 1860 Dade County had just eighty-three residents, and as late as 1920 it held only 5 percent of the state's population.[11] However, South Florida experienced a massive population increase and urban development during a series of property booms, most notably and consistently since the 1960s, as first migrants from Cuba but then also from Haiti and the Dominican Republic, among other Latin American sending countries, began arriving in numbers. Nijman describes the period from the 1960s to the 1980s this way: "It had been an amazing ethnic and cultural overhaul. Within less than twenty-five years, Miami's population had changed beyond recognition. . . . Miami turned itself 'magically' into an entirely new kind of multicultural metropolis or . . . 'Latin city.'"[12] Such an intense population growth and the demands it placed on infrastructure and services of all sorts proved a stimulus to the growth management measures that both local and state government sought to put in place from the 1970s. However, the population increase continued in South Florida into the 1990s, so that "while growth management in Florida may have managed growth, it did not limit it. Quite the contrary, the state's population and employment growth continued totally unabated during the 1990s. . . . The 1990s were yet another decade of unbalanced growth in a state still coming to terms with the 3 million net new residents that had arrived each of the previous three decades."[13]

South Florida was transformed during this period in a manner that corresponds closely to Harvey Molotch's description of the city as a growth machine.[14] In Miami, a growth machine dynamic was established before the 1980s that was unimpeded by public controls or interference.[15] The massive population growth experienced through migration and natural increase and the transience of that same population meant that a fragmented and particularistic politics and a bureaucratic local government consisting of fiefdoms provided poor regulation of development and created a propitious landscape for private land speculation and development interests to flourish without hindrance. In this way, "local government was constrained neither by business requirements of efficiency nor by democratic demands of voters," Nijman writes[16] Nijman's excellent account of the development of Miami does not stray into the story of its surrounding South Florida suburbs, where most of the population growth and development during this period were concentrated, and it is arguably the case that the growth machine dynamic in America has been most evident in new suburbs, that is, on tracts of unincorporated land in once rural counties. Dade was one such rural county that went on to accommodate much of the population growth and associated residential and other development from the 1960s, having to rapidly acquire the bureaucracy and other trappings of a metropolitan county that was formed in a union with the city of Miami. Much of this suburban development in Miami-Dade County ended up in an area known as Kendall. According to the Miami historian Paul S. George, "While the city of Miami has prided itself, since its inception, on having risen out of a wilderness to instant city status in 1896, Kendall can make a similar claim one century later, for explosive growth has become its dominant theme."[17]

The Growth of the Kendall Suburbs

The Kendall suburbs take their name from the British colonial estate manager—Henry John Boughton Kendall—of lands sold to Henry James Reed by the state of Florida in 1883.[18] George argues that Kendall possesses each of the major components of an "edge city" and is more or less self-contained. This rather contradicts other reports.[19] Perhaps more accurately, he notes that "in the public mind, Kendall represents a vast sprawling unincorporated area in the southwest environs of Miami-Dade County that has undergone explosive growth in the past generation. This region of hundreds of thousands of residents . . . stretches, in the public perception, ever more closely toward the eastern border of the Everglades."[20]

The growth of the Kendall suburbs in the 1960s and 1970s was spectacular and of itself did much to prompt the formation of the metropolitan Miami-Dade County in 1957. Moreover, the urban development that has taken place in the county has been viewed as emblematic of the problems facing the rest of the state more generally. For a start, almost all of the development in this formerly rural county has taken place in the automobile era.[21]

The history of the development of Kendall is one long familiar in Florida. It involved land speculation on the part of numerous small investors, but also larger corporate interests—an occupation celebrated, for example, in the self-published how-to manual, Cliff R. Leonard's *License to Steal*.[22] One company that was instrumental in many ways to the opening up of the Kendall area to the inexorable development of suburban housing tracts and associated developments was the Arvida Corporation. This was a venture established in the 1950s by one of America's wealthiest industrialists of the time, Arthur Vining Davis. Arvida was perhaps most notable as the pioneer of the condominium concept and associated ordinances that Florida is more or less synonymous with.[23] Perhaps less well known is the major role the company also played in suburban development in South Florida, and notably in the Kendall area of Miami-Dade County during the 1960s and 1970s. According to one reminiscence, Davis had bought extensive tracts of land in the Bahamas, Cuba, and Florida. Much of the Florida land was agricultural, and Davis apparently had grand ambitions to grow fruit. However, when the soil was found to be less productive for fruit, his attentions turned to development and to selling parts of the land.[24]

In a pattern that has been prominent in the fashioning of America's outer suburbs, proposals for retail development actually preceded major residential development—so much so that Dadeland Mall acquired and retained for some time the predictable moniker "deadland." Dadeland Mall was originally envisaged by Davis and the Arvida Corporation in late 1950s for the highly accessible location where the Palmetto Bypass, now Expressway, joins the Dixie Highway (US 1) on the basis of future growth in population, though the east-west road North Kendall Drive (on which most of this future population growth would depend) had not yet been extended to Krome Avenue, which marks the edge of the Everglades. The open-air mall was completed somewhat later, in the mid-1960s, by Meyerhoff Corporation of Baltimore, and was later expanded and enclosed.

In one of those quirks of history, the Dadeland Mall development was originally envisioned as a suburban city with high-rise apartments—an original vision not much different from what the Kendall Downtown area plan sought to reinstate some fifty years later. The plans consisted of a "complicated series of agreements" under which Federated Stores leased 140 acres from Arvida and handed over half to Meyerhoff to develop a retail mall. The remainder of the land was slated for a theater, automobile showrooms, a bowling alley, and restaurants south of the mall area. The original plan envisaged three phases of development, of which the shopping center was to be the first, but also a "complementary business center and apartment area." Meyerhoff's vice president Jack Pearlstone was quoted as suggesting that "in theory . . . when the area is completely developed, a person could live at Dadeland without ever having to leave for anything."[25]

As Davis's initial idea of fruit production receded, the enormous tracts of land that he had acquired in Dade County in the 1950s had, in less than a decade, begun to represent a potentially valuable asset as population began to grow steadily in the Miami area. To an extent, "Dadeland spurred the most expansionist era in Kendall history—an era that continues today as construction grinds westwards toward then 'glades.' Nice clean development after nice clean development, shopping center after shopping center, hopscotching out the main drag, North Kendall Drive, each seeming to spring up overnight where yesterday there was only farmland."[26] However, this rather understates the concerted nature of action among the various landed interests active in the county at the time. For Arvida, a major problem was that its land holdings were scattered across a swath of territory west of US 1 that was not served by roads and other infrastructure needed for profitable development. Thus Arvida was instrumental both directly and indirectly in unlocking the development potential of its own land holdings and those of others.[27] Arvida had earlier donated land for the Palmetto Expressway junction with North Kendall Drive near Dadeland Mall. As recounted in one local history, "Mr. Davis donated all the right of way to four lane North Kendall Drive west to Krome Avenue with the condition that the construction be done at once. Although at this time it was a road that led to nowhere, it was too good a deal for the County to turn down."[28] In fact, the extension of North Kendall Drive also required donations of land from numerous other landowners in the area and so set in train an initiative by a local attorney (with his own land holdings) to complete the right of way to Krome Avenue to the west.[29] This was to be the infrastructural spine for urban development to proceed rapidly westward throughout a vast area now known as Kendall. "By the late 1970s, Kendall claimed more highways and expressways converging within its environs than any other area of Dade County . . . setting the stage for the staggering growth that continues to define that region today."[30]

A Downtown for Kendall

Sitting along the I-95–US 1 north-south corridor identified for greater density of development at the state level, the East Kendall suburbs represent one of the very few concerted attempts to fashion a greater density of development and urban amenity. The Kendall Downtown experiment has already been hailed as a success of New Urbanism.[31] It is considered part of a broader trend whereby "an increasing number of Florida cities and counties are confronting the problems of sprawl—traffic jams, lifeless downtowns, environmental degradation and sheer ugliness—and are seeking solutions in new urbanism. . . . New urbanism has made significant inroads with both private developers and public officials, as Florida recognizes the negative impact of sprawl and looks for ways to build more attractive, livable, economically and environmentally sustainable places."[32]

Our story begins not with New Urbanist architects, nor with county planners, and certainly not with citizens, but with the private sector and the Miami-Dade Chamber of Commerce (South), which was instrumental in posing the question of whether the Kendall suburbs should have a downtown.[33] The problem was that "while the area has all the makings of a successful downtown—scores of prospering stores, office buildings, a hotel, even two metrorail stations it's jumbled together into a 60s-style agglomeration that, while big on parking lots and wide lanes, is short on amenity."[34] As a result, the Chamber South president of the time was quoted as asking, "If we were going to have a fireworks display in Kendall, or a parade—where would they go?"[35] Focusing on the mall area and associated commercial development at the junction of US 1 and metrorail, the Palmetto Expressway and Kendall Drive (see figure 5.1), the Chamber of Commerce took up this idea and promoted it to the county transportation and planning committees. By 1998 the initiative had evolved into a separate entity in the form of a Downtown Kendall Committee and a charrette exercise facilitated

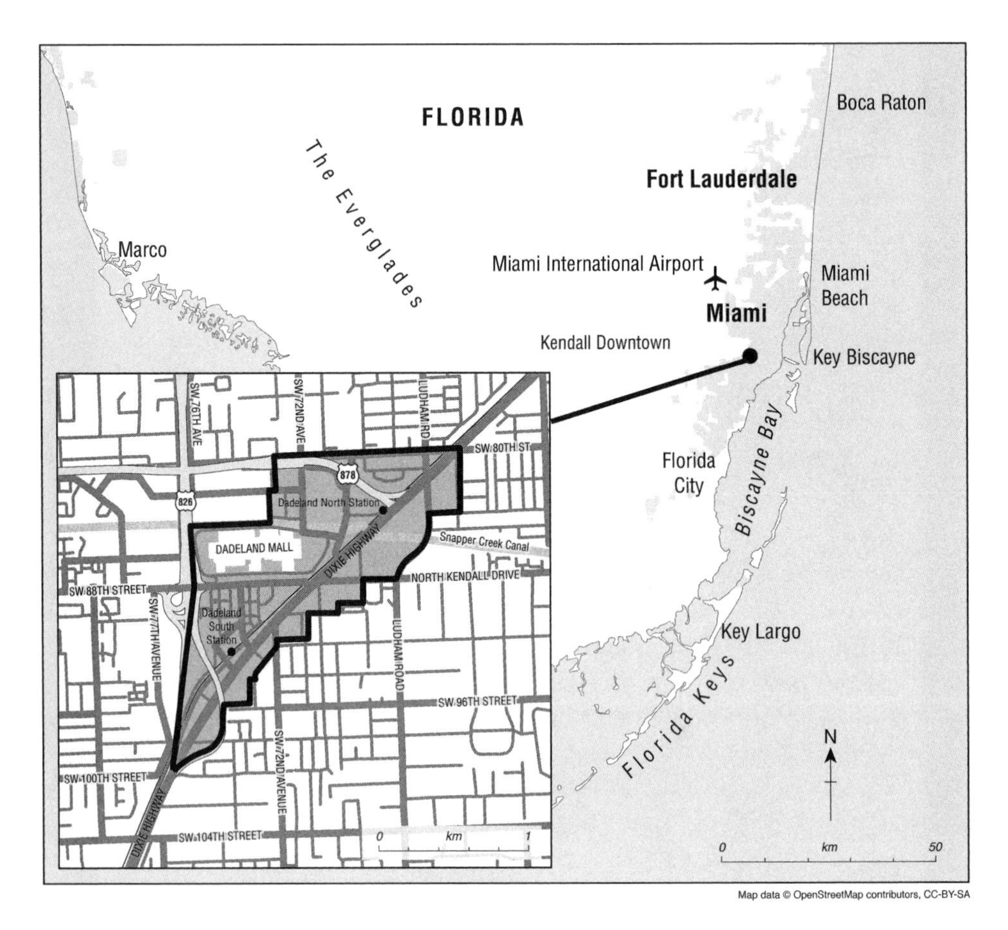

Map data © OpenStreetMap contributors, CC-BY-SA

Figure 5.1
The Kendall Downtown Area in Metropolitan Miami Context

by the Chamber of Commerce and involving local architects, planners, and citizens.[36]

It is an irony that the sprawl-plus environment of Miami-Dade County is nevertheless home to some of the key architects of the New Urbanism movement. Part of the New Urbanist movement, Dover, Kohl architects were instrumental in the charrette process for the Kendall Downtown area. Miami-Dade County planners were sympathetic to New Urbanist ideas, while nearby the city of Miami was also on a path to experimenting with "form based codes."[37] Thus, the New Urbanist principles of facilitating TOD and increased walkability quickly found their way into in the

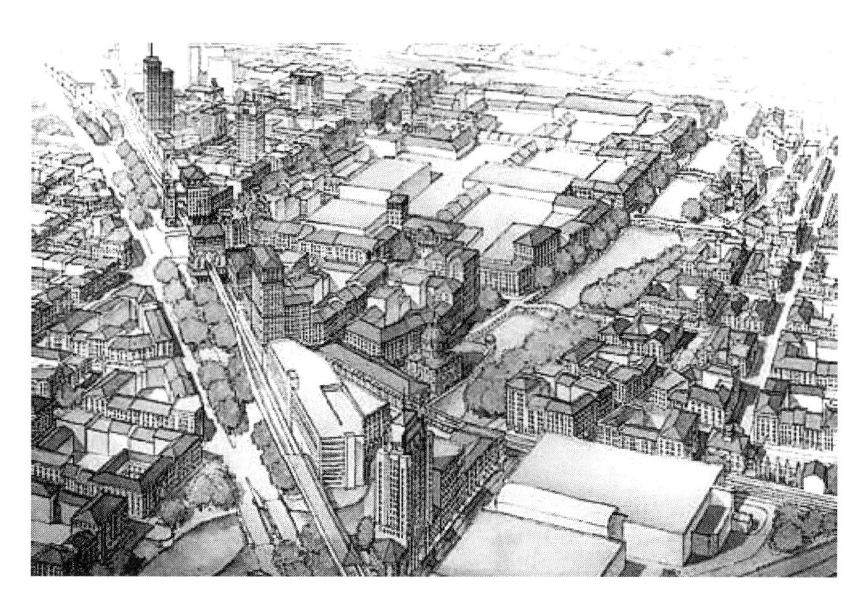

Figure 5.2
Architect's Sketch for the Kendall Downtown Area Charrette. *Source:* Reproduced with permission of Dover, Kohl Architects.

charrette exercise. They also crystalized within a newly created form-based zoning code for the Kendall Downtown area that was adopted in 1999 but subsequently altered in the face of a legal challenge by the Dadeland Mall owners, not least because it originally included the ideas of creating a grid of streets through, as well as developing the parking lots associated with, the mall.

The architect's sketch of how Kendall Downtown could look presented a seductive vision (see figure 5.2), and the complaint since among residents has been that "we thought we were going to get Paris and you gave us Manhattan."[38] For their part, the architects involved argued that "if we had not made it look so nice we would have got a lot less."[39] The fact that we are able to talk about the Kendall Downtown in the past tense indicates a certain measure of success. Thus, for one interviewee involved while at the Chamber of Commerce and since the mayor of a nearby municipality, it had been "extremely successful on a number of measures. A lot of times you do charrettes and they wind up as a plan that sits on a shelf and they never materialize, and that wasn't the case with us. People built things and that was pretty wonderful. . . . If you measure it on do you like this versus Paris, it is hard to hold up."[40] It is also clear that it is a work in progress. In this respect, according to another interviewee involved from the developer community, the ordinance has continued to be adapted and massaged to allow things to happen despite the original rigidity of New Urbanist precepts. "The vision is for it to be a city. It is not a city yet but it is getting there. . . . It has taken a while for all these condos to fill up and for people to get used to being in this area. There are other things that could be done. You could have a theater . . . and there are other uses, which is the beauty of this ordinance."[41] The character of the area has changed from almost totally commercial to 40 percent residential as a result of the development that has taken place on the back of the planning concept developed for the downtown and its accompanying zoning ordinance. In light of this it has been suggested that "I think what you see is an incredible model of what can be done. It is almost David against Goliath."[42]

For a former county commissioner involved at the time, whether the Kendall Downtown area can act as a model for elsewhere in the county is less clear. "I think it is too early to tell because of the lag in the economy. I think we need another ten years and things come back a little and see if people actually move there. And I also think it depends on if we are able to hold the urban development boundary or not. If we go back to the same kind of sprawling housing development eating up the farmland eating up environmental lands and if the builders prevail in that way, then I don't think it will be successful."[43] In this respect, the Kendall Downtown experiment appears to have failed, as yet, to provide much of a demonstration to the South Florida and Miami-Dade County property development market. Instead, as one developer noted, "Everybody has been looking for joint development along that path. There haven't been any other joint developments that I'm aware of with the county that have been successful."[44] In part, this reflects the approach to planning and the use of zoning ordinances, so that one interviewee questioned, "Why do we make it so that we have to have variances to do better things than to do worse things? There is a path of least resistance in most parts of the County that is to do the wrong thing and legally. . . . One of the things that I hope places like downtown Kendall will do is that they will demonstrate to people alternatives in the market. . . . So hopefully some of these things will start doing well in the market."[45] To date, it appears that the finance and development communities have not been especially convinced. There has been extremely limited TOD development at other stations along the north-south (Orange) line metrorail corridor to date, reflecting the limited appetite for such development across America at present, but also some of the peculiarities of the South Florida property market. If transience is perhaps the signature feature of Miami, this also seems to have affected the prospects for the likes of TOD and New Urbanist projects elsewhere in the suburbs of Miami-Dade County. Thus as one property developer with the biggest TOD-style development at Kendall Downtown explained, "You know, getting financing on a ninety-nine year lease is always a challenge. There is always just a lot of negotiations that goes on in developing a lease for that kind of term and to meet the goals that the county is trying to meet and for you to meet your goals to get it financed and built, and that process can get very lengthy and onerous, and sometimes it just fails because it takes so much time to get it done, the market changes. Especially here, you know—one day it is office, the next day it is rentals and condos."[46] This interviewee noted the speed with which developers in South Florida and Miami-Dade have typically switched from one segment of the market to another as individual segments of the property market get full.

Eastward Ho! Urban Sprawl Meets the River of Grass

Florida's approach to the question of growth management—the management of this tension between the urban development that accompanies population and economic growth and its impact on the environment in terms of land take and environmental impacts of all sorts—has been considered unique among states. The adoption, regular review, and examination of comprehensive land-use planning by counties and cities by the state have been at the heart of this approach, which has also sought "concurrency" in the planning process and government budgeting to ensure that individual approved developments do not take place without the associated requisite infrastructure and a measure of dialogue between local governments.[47] Indeed, "the Florida approach is generally acknowledged as the most aggressive and far-reaching growth management approach this nation has yet seen."[48]

Growth Management or a New Frontier for the Growth Machine?

If Florida identified comprehensive planning as the strategy for managing growth, it is acknowledged as having failed.[49] Comprehensive planning as a tool for growth management does not function in the manner originally intended or designed—not least because the state's own behavior undermines concurrency and compact urban form.[50] Instead, "when push comes to shove in Florida, and growth management and environmental protection are perceived as a threat to economic growth and prosperity, Florida's legislature has traditionally taken steps to undermine growth management and environmental protection."[51] Perhaps the ultimate proof of this underlying bias toward growth over the environment was provided by the recent repeal of Florida's growth management act with the desire to stimulate an expected 700,000 jobs across the state in response to the economic downturn since the subprime mortgage–induced financial crisis of 2008.[52]

Statewide anxieties over growth and environment have coalesced in recent decades under the slogan Eastward Ho! That is, there is desire to prevent urban sprawl from eating any further into the river of grass that is the Everglades' sensitive ecosystem.[53] The state governor's Commission on a Sustainable Florida of 1995 attacked suburban sprawl. "Growth has also given rise to a proliferation of low density development and other negative consequences including gridlock, isolation, disparity, factionalism, mind numbing homogeneity, and a distinct lack of a sense of place."[54] The I-95 and US 1 corridor emerged from these state deliberations as one designated for an increase in the density of development. Despite what is regarded as the denucleation of South Florida, Robyne S. Turner and Margaret S. Murray suggest that there are clusters of communities in the I-95 corridor that have not been recognized.[55] However appealing greater development density may be in technical, planning, architectural, and engineering terms, and indeed in terms of contributing to greater environmental sustainability, it is also a politically charged policy because of its likely displacement effects on lower- and middle-income families and on particular ethnic groups. In Miami, these displacement effects take on a greater significance as a result of the mistrust of government institutions that has persisted after urban freeway construction projects decimated the once vibrant black neighborhood of Overtown.[56]

Turner and Murray have argued that "under the title Eastward Ho! Politicians and developers alike are looking to the area as the next frontier of development and increased tax revenues."[57] Doubtless greater density can present additional profits for developers, especially in mature urban environments where land and property have greater use value; however, the Kendall Downtown case also reveals the limited appeal of this frontier to extant private corporate and residential interests.

The elephant in the room regarding the attempts to plan for a more dense and mixed-use downtown for the Kendall suburbs was the enormous Dadeland Mall (see figure 5.3, where Dadeland Mall lies to the right of the metrorail). The mall had been pivotal in the development of the rest of the suburbs and was, as we saw earlier, originally envisioned as the core of a set of high-rise developments. However, despite the discussion of the potential for retrofitting or redeveloping dying or dead malls, there is as yet little evidence that the owners and operators of suburban shopping malls perceive any of the limits of the suburban retail market. The original vision of the architects and planners who participated in the Kendall Downtown area charrette was that the frontage of the mall could be adapted and improved and a grid of pedestrian streets could be opened through this enclosed space. There is a paradox, then. On the one hand, the mall was the primary and most powerful objector to the scheme. As one interviewee,

Figure 5.3
The View South toward the Kendall Downtown Are. *Source:* Author's photograph.

a former county commissioner, said of the planning process and the adoption of a zoning ordinance covering the Kendall Downtown area: "On the bright side, the whole mixed-use concept has worked out reasonably well. The idea for the shopping center was that eventually it would be much more connected. The shopping center was one of the more difficult partners in the whole thing because [the owners] wanted to retain their sea of parking and they didn't want to build up their buildings right to the edge. They wanted to keep their same style. They were very uncompromising."[58] A late legal case was brought by the mall owners against the new zoning ordinance with a particular eye to any likelihood of a grid of streets being worked through the mall or across its expanses of car parking space. "Both sides broke bread during the county-funded planning workshop in 1998. But a schism formed after zoning changes became official, prompting Dadeland to file a formal claim of property rights violation."[59] Yet on the other hand, according to a representative of one of the main developers involved: "What made this go was the mall. So unless you have a mall or something that is a driver, I don't see how a city could follow around it."[60]

Some of the appeal of the density envisioned in the Kendall-Downtown planning exercise was the product of the booming property market at the time and as such one of the reasons for Kendall Downtown's relative success was simply the timing. Yet, the propitious timing of the Kendall-Downtown example is also curious in that it reveals the limits of interest in and even the expertise of developers in mixed-use, dense developments. Lennar Corporation, a major suburban house builder, became interested in the development, though it is acknowledged that the corporation probably would not choose to do so again after having its fingers burned when it was unable to sell apartments that were overspecified.[61] One developer went bankrupt after considering it a sound idea to excavate underground car parks into the limestone rock with its shallow water table beneath the showpiece arcades development. For some, projects such as Kendall Downtown have a value—alongside a number of comparative studies of the economic returns of typical suburban versus urban development formats—in providing an important demonstration of the potential of New Urbanism to effect change across American metropolitan areas. Yet there is also some recognition otherwise: "Still, the lending and financing industry—they do not understand properly the benefits of a long-term investment in a good urbanism."[62]

The lack of appeal of urbanizing suburbia to proudly suburban residents in a metro area considered one of the most centerless in America also can hardly be ignored. As one of the pioneers of New Urbanism explained, "One of things you have to deal with in suburban and edge city retrofit is respect the single suburban house. There is a certain kind of power that exists because they are voters and you need their support because they are the worst NIMBYs and they can prevent things from happening. Suburban redevelopment is not about redeveloping single-family units; that is sacrosanct, and you want them to support you in whatever else you are doing."[63] For another New Urbanist architect and author, residential areas are likely to be the last thing to go in any trend toward a greater density of development in the suburbs. Moreover, residential redevelopment at greater densities, perhaps in line with New Urbanist principles, may only really be possible if it is regeneration of blighted areas affecting potentially poorer residential neighborhoods.[64] As the same architect explained, "When the pressures of location and land costs converge, at some point there may be transformation of the residential communities."[65] As Arthur Nelson has cataloged, there is ample suburban space that is not residential and could provide the basis for a greater density of development in suburbs.[66] Yet it is also true that the sacrosanct nature of residential areas presents some real problems to maximizing the potential for TOD and for New Urbanist projects to effect a significant transformation of suburbia.

Coincidentally, at the time the Kendall Downtown proposals were coming forward the wealthy citizens of Pinecrest, immediately adjacent, were incorporating themselves as a distinct suburban community. Pinecrest is an exclusive large-lot, sidewalk-free, primarily residential community bordering on its eastern limit the Florida coast and on the west the US 1 Miami metro rail corridor, seen as the driver of greater density at Kendall Downtown. This immediately presented a powerful and separate jurisdictional opposition where there had previously not been one. The Datran Center (see figure 5.4) had already been completed before the incorporation of Pinecrest. While incorporation, coupled with existing opposition from Pinecrest residents, undoubtedly has ensured a lack of corresponding density to the east of US 1 and the metrorail (see the contrast between Pinecrest to the left and the Kendall Downtown in the center of figure 5.3), it was thought not to have been a major source of opposition to the Kendall Downtown proposals as a whole. However, it did prompt debate regarding the possibilities of incorporation

Figure 5.4
View of the Kendall Downtown Area from Pinecrest. *Source:* Author's photograph.

among residents of the 1950s Continental Park, who have been concerned about the potential of high-rise development to jump the western boundary of the Kendall Downtown area.[67]

Residents of Continental Park on unincorporated county land west of the Kendall Downtown scheme were concerned about any creep westward of the greater densities envisioned. They wanted the greater density of development anticipated by planners and architects contained east of the Palmetto Expressway. Here again, residents proved some of the most difficult constituents to be persuaded of the merits of greater density, though this was in any case less critical to the efficacy of TOD development than was the opposition in Pinecrest.

The Elusive Search for Concurrency in Miami-Dade County

"Despite the progressive leadership [at state level] . . . Florida's demographics changed so dramatically in the 1970s and 1980s that the state's population growth increasingly overwhelmed most environmental reform initiatives."[68] Instead, as David Colburn and Lance deHaven Smith go on to note "Growth has placed a heavy demand on state and local governments to provide roads, water and sewer systems, schools, and other public facilities. But Florida's tax structure is such that state leaders cannot meet this challenge without constantly going back to voters for increases in the tax rate. And Floridians have typically rejected such requests in the past two decades."[69]

Growth Management and Parasitic Suburbanization

Since "nearly every major aspect of Florida's problem of controlling growth and protecting environmental quality can be found in Dade County," "policies for controlling land use and guiding growth are not likely to succeed in the rest of Florida if they do not succeed in Dade County."[70] Indeed, growth management could be said to be at the heart of the formation of Miami-Dade County as a metropolitan authority, while the process of catchup, of supplying infrastructure and service and amenities for population and residential growth, has hung over the county until the present day. Miami-Dade County's early policies governing land use tended to be indiscriminantly pro-development. In particular, L. J. Carter argues, commissioners were lax in living up to the concurrency idea within the adopted General Land Use Master Plan.[71]

During these years, "The more persistent, unremitting political pressure on metro [Miami-Dade County] came from people with a stake in either the buying or development of land—landowners, contractors, labor unionists in the building trades, local mortgage lenders, zoning lawyers, and numerous outside investors, including major eastern banks, insurance companies, and industrial corporations."[72] Nevertheless, the Comprehensive Development Master Plan (CDMP) was successfully put in place but against this backdrop, in which "it took the County over a decade to control a semblance of unified government services as it faced tremendous pressures of lack of resources, continual population growth, a new dedicated concern to protect the natural environment, and a concern for appropriate zoning and land use controls."[73] The new CDMP was adopted in 1975 and revised in 1978, and represented an ongoing attempt to manage growth positively toward a set of outcomes that involved greater density of development, transit orientation, and concurrency, since "zoning is particularly deficient in 'making things happen,' although it can allow things to happen."[74] However, the plan also noted that in this respect, "Dade County does not have adequate tools to guide its future growth and development."[75] During this time, some of the first refusals to allow housing development had been apparent, based on the lack of adequate provision of services and infrastructure and subsequent legal challenges. "For the first time in metro history, the County commission rejected a proposed housing development in West Kendall solely because there was no plan for school construction to accommodate the projected number of children."[76]

One controversial planning tool put in place by Miami-Dade County over this time was an urban development boundary (UDB) in 1975 enveloping 364 square miles. The UDB was formally instituted in 1983 in the county's CMDP and has been expanded over the years by 15 percent to include 420 square miles.[77] In theory, among its many potential effects is an intensification of development within the UDB as a result of displacement of demand to, and rising land prices within, the UDB.[78] The UDB is intended to protect and preserve the wetlands of the Everglades National Park to the west, prevent low-density development away from transit and neighborhood amenities and unconnected development patterns, and plan for the efficient expansion and improvement of infrastructure and public services. However, the UDB has proved to be a lightning rod for the expression of conflicting views among politicians, planners, the private sector, and citizens and environmental groups. It has been subject to constant political lobbying and legal pressure from developers for parcels of land to be included within it in a pattern that might be seen to reflect the growth machine politics of business interests in the county and the state (see figure 5.5).

In reality, relatively few applications to enlarge the area included in the UDB have been approved since the adoption of a two-thirds majority rule instigated in the 1990s.[79] However, the UDB is likely to come under renewed and intensified pressure in the near future. Miami-Dade County is projected to have a population increase of 700,000 residents by 2030, with the UDB intended to ensure a fifteen-year supply of land for development. The problem is that "the remaining land inside the boundary may not yield a 10- to 15-year supply unless the county, in cooperation with the 35 cities, promotes infill development on vacant and underused land."[80] In this respect, the review suggests that the infill development record among cities

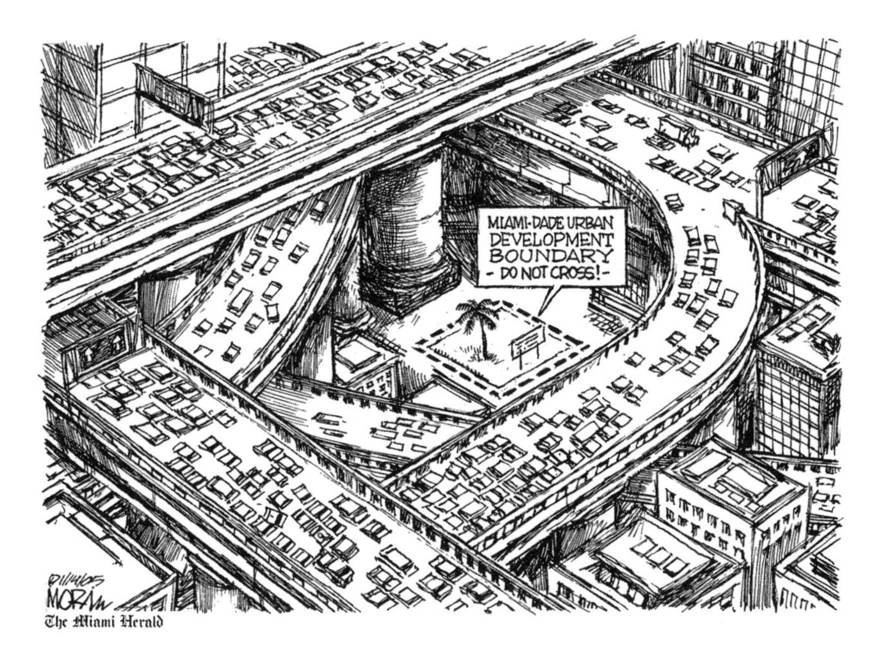

Figure 5.5
Miami Herald Cartoon Lampooning the Pressure on the UDB. *Source:* Reproduced with permission from Jim Morin.

is decent compared to elsewhere in Florida, but that "perhaps the most important area to address is how the county can coordinate infill development with the thirty-five cities in its jurisdiction. Much of the available land inside the UDB is inside city boundaries. In theory, the county has the power to assert land-use control over critical areas inside cities, as it has done around transit stations. But asserting further control, though possible, is fraught with potential political problems," not least, as we saw earlier, that of persuading constituents in incorporated, primarily affluent residential suburban "cities" of the merits of urban infill.[81]

A second planning tool that emerged in later iterations of the CMDP in the 1980s in Miami-Dade County planning was the concept of traditional neighborhood developments (TNDs). "The purpose of a [TND] is to enable the creation of new communities that offer social and architectural quality, characteristics of early American town planning."[82] Despite adhering to principles of traditional architecture and urban design and contributing to a lessening of some of the more excessive separation of land uses, this policy has been less successful in promoting greater density and TOD. Instead it may have made its own distinctive contribution to the dysfunctional planning for mobility and the unique sprawling and centerless appearance of identical blocks of development across much of Miami-Dade County and notably the Kendall suburbs. One interviewee complained that this pattern of development revealed a lack of hierarchy in roads planning and responsibilities, little in the way of a culture of developer contributions, and often little thought given to preserving rights-of-way ahead of the developments.[83]

A third planning tool that has proved controversial and sluggish in its response to growth pressures is the attempt to marry comprehensive planning objectives to the provision of mass transit infrastructure. The 1975 plan described how "the county's policy attempts to accommodate the area's automobile population increase is being challenged on two fronts. First, in the urbanizing areas of the county, roads cannot be built fast enough to keep up with the increased demands. Second, in the already urbanized areas, local opposition is preventing the construction of more expressways and some major arterials."[84] As the subsequent revision of the plan went on to elaborate, "Dade County's rapid population growth and dispersed pattern of urbanization has severely tested the ability of its transportation system, especially its road network, to provide adequate levels of accessibility for intra-urban travel. The low density development presently occurring in Dade County's urban fringe area has been made possible in part by the extension of several arterials into formerly undeveloped locations."[85] In response, the plan placed emphasis on the ability for a planned heavy rail public transit network to deliver greater development densities. "Public or mass transportation shall be given top priority as a positive tool to support and improve the viability of the county and the region."[86] In the run-up to the referendum, the Miami-Dade County manager argued that "we are going to sprawl and the cost of delivering services, water, fire, police, schools, roads will be tremendous. Taxes will go up and the quality of life down."[87] "Activity centers" were identified in the plans of the 1970s as those around which the metropolitan economy could be structured, while later plans also identified "major activity centers" (of regional, South Florida impact), which would drive efforts to promote a greater density and transit orientation of development.

"Metrofail" and the Limits of TOD

However, provision for mass transit in Miami-Dade Countyhas been plagued by problems from the outset such that it has acquired the predictable moniker "metrofail." A recent review surely puts some gloss on the case when it argues that "metrorail transit system and planned expansions of both rail and bus rapid transit have created a solid backbone upon which to create more compact nodes inside the UDB. The County has pursued this type of development in unincorporated urban centers such as Kendall and Naranja."[88] The potential for transit to drive such opportunities has been extremely limited as the metrorail system has existed for much of its time since the 1980s as a single north-south line, only acquiring a spur to the airport in 2012. It originally consisted of ambitious plans for several lines, requiring a vote on an addition to the sales tax to be levied to fund it, which was closely fought. The provision for mass transit in Miami-Dade County has been plagued by problems from the outset as a result of several interrelated peculiarities of the Sun Belt urbanization and politics and government in this part of the country.

For one interviewee, a regional planner, the problems surrounding the development of metrorail as a collective infrastructure to drive a greater density of development reflect a fundamental conundrum faced across much of America in terms of the financial case for mass transit, but perhaps more so in those Sun Belt metropolitan areas where growth has been of very recent vintage. "It's the chicken versus the egg. There isn't sufficient infrastructure . . . but then you don't have the densities to build the infrastructure. . . . Florida is different to the older cities where the infrastructure predated the development. So now it is a matter of retrofitting transit around development."[89] The same point was amplified by the representative of one developer involved in the Kendall Downtown planning experiment but who nevertheless saw eventual benefits of the limited development of metrotrail to the developer's own corporate interests. "The whole idea of this is that people would use mass transit. But the citizens of Dade County are not like the ones in Washington, D.C., or New York or Boston. They are not used to using a train to commute. They like their car, but the roads are getting packed. There are 60,000 new residents coming in to this area each year. They have got to live someplace, and they are driving their car. So eventually this is going to work even better because people are going to have to use transit."[90]

Metrorail's problems as a collective solution to the mobility needs of Miami-Dade County residents reflects the character of the state politics. As one former county commissioner said, "There are lots of plans but there is no money. The biggest problem is that we have a very anti-tax community right now. We were able to pass a half cent sales tax for transit but it was not just for transit it was for transit, roads, sidewalks. . . . People felt as though they had been hoodwinked. It was sold as a transit tax and they were getting all this other stuff and now there wasn't enough money for transit. What was promised was enhancement of projects, not the basic transit system. What we found was that we needed the money to prop up the basic transit system and our buses. So people felt so betrayed by that."[91] The same account is amplified by a representative of the Metropolitan Expressway Authority, who noted that the failure to deliver a more extensive metrorail network during this time left an urgent need for further expressway building to serve traffic growth. "Herein lies the Miami-Dade problem the way I see it. Twelve years ago the former mayor of Miami-Dade County put in a 'taxes for tolls' initiative—meaning get rid of all tolls in the region, and we are going to tax you instead—and it failed miserably. People said I would rather pay for my roads when I use them. So a couple of years later in 2002 they started this half penny sales tax and said I am going to give you a north line, a middle line to the west, plus more buses and all these things. About 2006 rolls around we don't have the money, we oversold what we could give you. Heads rolled at transit and the mayor has gone. . . . Then, to compound that, over the last ten years at the same time all this growth has been happening. You get to the point where 'do nothing' is not an option anymore."[92]

If provision of road infrastructure for the use of private vehicles and provision for the collective consumption of mass transit have been to a large extent opposed in the historical development of America's postwar suburbs, then this opposition falls into sharper relief in the Miami-Dade area. For a commentator from the roads agency, the noble aims also proved a problem in connection with parallel efforts to provide for the use of the private automobile. "Something interesting happened in Miami back in the 1960s, 1970s and 1980s. You had the department of planning and zoning, [which] was trying to control growth and didn't provide transportation avenues. They tried to prevent sprawl but sprawl happened. From here all the way south is completely built out, and you can see there are no expressways there, there is nothing to get you out of there."[93] The problem is compounded by the fact that the different county institutions have come to be regarded as something of bastions for particular ethnic and class interests and consequently are pitted against each other for funding.[94]

Governing Post-Suburbia: The Two-Tier Metropolitan County Solution

"Because it is such a new city, and its growth has been so rapid and so recent, Miami is characterized by 'suburban sprawl'— a relatively small urban core surrounded by a . . . built up area in which one town runs into another with no particular community focus and little differentiation in appearance or function."[95] With a few exceptions, this perceptive picture painted by Reinhold P. Wolff over half a century ago remains remarkably familiar today. While the Brickell area in particular has grown upward, with office and condominium towers, as an addition to the downtown, the view from downtown Miami presents a largely flat horizon of trees under which single-family homes and low-rise apartments and townhouses spread for miles, punctuated by a few small spikes of density. Yet metropolitan solutions have foundered on the peculiar nature of population growth and its effects.

The Transitory Suburb and Metropolitan Government

The extent to which urban development in Florida and Miami has been fueled by in-migration has meant that suburbs have taken on a rather different complexion here than elsewhere. Wolff again: "From the very beginning, the suburbs in southeastern Florida played a different role in the area's economy than that of most of their counterparts in the north. Started by their original developers as independent and self-contained municipalities rather than as suburbs, cities in Dade County are in many respects not comparable to the rural towns that in the north grew into the orbit of the expanding metropolis, nor are they typically 'dormitory towns' to which the city dwellers drifted in search of more comfortable small town living. Most of Miami's suburbs were founded by migrants from other regions who hoped to create fully integrated cities in which people not only lived or slept, but where they found a livelihood as well. Thus they were not really 'suburbs' in the first place."[96] This was a picture elaborated shortly afterward, in 1963, by Edward Sofen: "Miami's suburbs do not appear to be typical of suburbs in most other metropolitan areas. In greater Miami the population is constantly shifting and growing, not with the overspill from the central city but with in-migration from all over the United States."[97]

Wolff talks of the denucleation of the suburbs and the loss of differentiation among them as a result of successive waves of migration. This was a characteristic that appears to have later reinforced support for the institution of metropolitan government in the creation of Miami-Dade County and has been a theme returned to in recent pronouncements from the state's governor. Thus, "In the Miami area such unincorporated nuclei are formed in great numbers but not all of them have sufficient distinctiveness or size to be identified as separate towns. They have contributed a great deal toward complicating the municipal structure. On the one hand they have generated a new civic concern of the small town type that is so characteristic of the spirit of suburbia. On the other hand they have tended to wipe out differences that once existed between incorporated towns and unincorporated areas. . . . They have tended to strengthen the desire for metropolitan rule."[98] The weakness of interests that might elsewhere have posed serious threats to metro government—including the organization of ethnic minorities at the time—presented less entrenched opposition to the idea of metropolitan government.[99] Yet they also undermined it in other respects. "The transitory nature of Florida's residents . . . has made it exceedingly difficult to construct a political consensus to address state and local needs."[100]

Moreover, the peculiar character of the South Florida economy has played into the sorts of pressures faced by metropolitan government. "While tourism is the backbone of the economy, Metropolitan Miami is a classic example of an area whose growth has been nurtured by the basic needs of its own residents."[101] These basic needs are those associated significantly with land development, and they have continued to present a challenge to county government and planning. "From the outset, Miami has been a case study of how not to reform, and many of the political wounds remain open today."[102] It is a two-tier metropolitan government solution unique in the state and nation. However, as the county's own revised CDMP declared, "While inter-governmental flexibility is inherent in these provisions, the failure of the Charter to resolve the issue of functional responsibilities set the stage for a continuing series of political and legal challenges."[103] It extends into suspicion over county government itself. "Even today, Metro is viewed with some ambivalence by city officials."[104] These suspicions extend to the county's different service delivery organizations. In particular, the popular perception is that particular bodies are understood to be fiefdoms of particular class and ethnic interests, while there is also a sense of competition among different bodies such as Metrorail and Metro Expressway Authority (MPX).

Who Owned Kendall Downtown?

The political and legal contestation between the county and the cities has been open in relation to issues of land-use planning, as we saw above in relation to the county's attempts to promote greater density of development. The idea of deploying the mass transit system to drive TOD was itself one issue over which the two-tier system became politicized since cities were unhappy with the county's desire to assume powers of zoning on transit impact zones surrounding stations.[105] Moreover, some of the full effects of TOD were curtailed by the objections of Pinecrest residents. Yet, as Elizabeth Plater-Zyberk has noted, "The fact that stuff has gotten done according to plan is quite a victory. But the plan always understood there would be no one choreographing it." She qualified this statement by noting that "the whole implementation aspect of it is interesting because most of these edge cities are in unincorporated counties which are now acting like municipal governments but don't have the kind of focus that a city government does."[106] And for others, though Kendall Downtown can make its contribution to a more walkable and livable suburb, there is no likelihood of it becoming a city.[107]

However, for another observer and resident activist, the eventual end-product of the project was compromised by the lack of a dedicated body to oversee the plan implementation. "This is unincorporated Dade County. The details were not gotten right every time in Coral Gables and South Miami but at least there was a Coral Gables and South Miami to watch over it. All that was left to watch over the details of this was a very powerful public works department, an inept transit department . . . and a not so great planning follow through."[108]

Curiously, some of the biggest challenges to the planning vision for Kendall Downtown area, a vision that was established through the charrette process and incorporated into a new zoning ordinance and the "not so great planning follow through," were the actions of other departments within the county government itself. These tend to underline some of the difficulties of fashioning a more urban mix, density, and format of development in ostensibly suburban government jurisdictions. One planner closely involved in the charrette process noted that the desire for greater pedestrian friendliness came up against the Department of Public Works' requirements for the positioning of ventilation outlets and fire hydrants. while the Highways Department actually widened North Kendall Drive to create a bigger divide between the Dadeland Mall and the pedestrian-friendly showpiece colonnade-lined Kendall Downtown development (see figure 5.6).[109]

Conclusion

The case of Kendall Downtown illustrates how "the 'new urbanism' . . . can make urban living more satisfying, but it cannot solve the state's enormous urban growth and resulting urban sprawl."[110] As a glimpse of the Kendall suburbs' post-suburban future, the Kendall Downtown development highlights how the county government's attempt to balance intergovernmental relations, provide for collective consumption, and promote both growth and the environment was overwhelmed by the sheer weight of the population growth and the associated pressure for residential development as these have appeared in a metropolis of very recent origin.

Strangely, the New Urbanism flowed out of Florida into a movement with something of a national, even international, appeal. If Miami has become a world city, this status is very recent and does not rest on a particularly firm basis. Miami's transience has tended to favor ad hoc urban projects at the expense of longer-term design and planning.[111] The Kendall Downtown area might be considered one such ad hoc project, though one that is informed by distinctly different principles than have been applied elsewhere in the metro area.

The Kendall Downtown area has been spoken of as a success story, and perhaps it can form something of a model for the rest of the metro area and beyond. Yet against this, "The suggestion that Miami will 'mature' over time ignores the fact that in this city transience and civic challenges moved in tandem for well over a century. Miami's character is built on transience."[112] This suggests it may well remain an isolated experiment surrounded by considerable uncertainty regarding its wider meaning to the public, planners, and the private sector in the metropolitan area. Some of

Figure 5.6
Across the North Kendall Drive Divide: The View toward the Kendall Downtown from Dadeland Mall. *Source:* Author's photograph.

the shortcomings of the Kendall Downtown experiment of itself, namely, its peculiar timing and its limited reception elsewhere in the metro area, ensure that "Miami will not be another Manhattan any time soon."[113] The Miami metro area may have diversity, but it does not have density, nor, judging by the Kendall Downtown story, is it likely to have density anytime soon. Instead, the story of Kendall Downtown is one of the limits of the New Urbanism in a Sun Belt setting. As one planner, sympathetic to the Kendall Downtown experiment but more critical of its wider meaning and significance, argued, "One of the frustrating things is that the new urbanism community . . . tends to hark back to their roots, which are northeastern, come down here and want to put New York City solutions on South Florida problems. New York City and Philadelphia, Chicago and Boston all developed in a pre–World War II era for the most part. The market doesn't develop like that anymore. It is certainly resisting it here. . . . The creativity is lacking to think about South Florida as South Florida. It is an expressway-oriented town. . . . It is probably more similar to Dallas or Los Angeles or Phoenix than it would be to New York City, yet no one comes with a Sun Belt solution."[114]

While powerful planning ideas are always born of a particular set of influences, always born of a particular time and place, it could be suggested that it is those planning ideas that are strangely timeless and rather placeless that have been the most powerful and have traveled the most extensively. If the Kendall Downtown experiment is already part of the past of New Urbanism, then what it may reveal is the limited wider impact this particular planning idea is likely to exert on the remaking of American suburbs with its often very specific appeal to historical reference points.

6 Taming Tysons: Post-Suburbia's Present?

The world is being used as a model for Tysons and we are trying to use Tysons as a model for the rest of the United States. . . . I see this as a problem all over the United States and if you can tackle Tysons and solve it you could solve anything in the country. . . . We felt we could help demonstrate how to repair the damage done to the environment and to society over the last fifty years with this attitude of just build and sprawl and so on.

—Partner, KGP Design Studio[1]

Tysons Corner is perhaps *the* archetypal edge city, and it is tempting to characterize its development as a product of spontaneous private-sector initiative.[2] However, it is every bit the product of the sorts of state interventions—including major road infrastructure and land-use planning activity—and their unanticipated side effects that were discussed in chapter 3. Even in one of the most pro-development state contexts in the United States, private-sector initiative in Tysons Corner and in suburban Fairfax County, where it is located, has taken its cue from, and has an important relationship to, state intervention and the planning process.

Indeed, Tysons Corner has been recognized in county planning exercises since the 1960 Plan for Industrial Development; was subject to a first Master Plan of Tysons Corner in 1963; and has been covered in several land-use plans and other visioning exercises since. Tysons Corner's future was foretold in the Maryland-National Capital Park and Planning Commission's general plan for the greater Washington, D.C., city-region, which depicted the location that is now Tysons Corner as a potential growth node along a series of axes.[3] However, as a result of the peculiarities of the capital city's metropolitan region, planning and intergovernmental relations at this scale have exerted very little effect on Tysons

Corner's emergence, nor are they likely to have much bearing on its future course of development.

In 2010, the deliberations of the Tysons Corner Land Use Task Force, instigated in 2005 by the Fairfax County Board of Supervisors, finally resulted in the adoption of a new comprehensive plan, replacing the existing plan, which dated from 1994 (with subsequent revisions). This latest planning exercise recently won the Daniel Burnham Prize from the American Planning Association. After a discussion that has incorporated many interests and many different potential scenarios, it proposes a significant reworking of Tysons Corner's suburban space into a proper downtown and represents something of a present-day test case for attempts to retrofit the very many out-of-city office and retail nodes across America.

For all this, Tysons has been a difficult beast to tame, and almost certainly will prove so in the near future. Its growth could be regarded as iconic of neoliberal suburbanism, in which suburban development has been a stealthy carrier of "subgovernance," with all that entails regarding the nature of the relationship between planning and state interventions, such as the role of mass transit in private accumulation.[4] Any upward growth of Tysons Corner may yet be a manifestation of the growth machine all over again.[5] Here, though, following the scheme presented in chapter 3, I describe the local politics surrounding the reworking of Tysons Corner as a community on the cusp of transitioning from growth machine–style to development regime–style politics.

As one of the very few examples of the extension of mass transit into the automobile-oriented outer suburbs of America, Tysons Corner is a live example to look to for evidence of the effects of the possibilities of transit-oriented development (TOD) to drive a reworking of the suburban way of life. As highlighted in the epigraph to this chapter, a quotation from a practicing architect involved in the deliberations on the future of Tysons Corner, it is a potentially seminal moment in the history of the outer suburbs, a moment in which America's post-suburban future may be present. Yet even in this moment in which a post-suburban future might be glimpsed, the new frontier of urbanization in America has already moved on to the yet more dispersed form of edgeless cities.[6] It is something to conjure with.

Tysons Corner and the Fairfax County Growth Machine

Tysons Corner existed as a rural crossroads until the 1960s. It lies along roads dating from colonial times, and, as one of the highest elevations around, it became a signaling post in the Civil War.[7] A communications tower, a modern-day remnant of this strategic position, still stands among the office buildings that now populate the location (see figure 6.1). Tysons is one ingredient in the restless urban landscape reflecting in turn important changes in the imperative for capital accumulation in the greater Washington, D.C., metropolitan area.[8] Curiously, however, even the evolution of the economic base of Tysons Corner and the larger Dulles Airport corridor into Internet-related software, services, and consulting activities appears not to have escaped these and other long-established locational attributes.[9]

The proximate origin of the initial speculation in greenfield land beginning in the 1960s and its conversion into urban land soon after can be

Figure 6.1
Communications Tower at Tysons Corner. *Source:* Author's photograph.

found in two rather different tributaries, one from without the community, in the form of significant federal and state expenditures, and one within the community, in the form of the early acquiescence of residents to a post-suburban politics of local sufficiency in place of the local liabilities associated with a purely residential community. One curiosity is that it is the private residential and commercial growth of Tysons Corner, unleashed by major public investment for the private use of the automobile, that now stands to be reworked again through public investment for the collective consumption represented by mass transit.

Tysons Corner and the Systematic Distortion of Regional Patterns of Accessibility

The regional accessibility of the location that is now Tysons Corner was massively enhanced by major expenditures by federal and state governments devoted to road infrastructure improvements. Notable among these improvements were the building of the Washington Dulles International Airport, an associated access road, and the Capital Beltway, but also expansions of state roads.

The handful of "growth entrepreneurs" that set development in motion realized, for different reasons, the locational attractions of Tysons Corner.[10] As the former chief executive of Fairfax County observed, it was not developers per se who developed Tysons but individuals who had acquired land and held on to it for a long time.[11] Prior to the construction of the Capital Beltway, roughly two-thirds of the primarily dairy farming and quarry land at Tysons Corner was in the hands of just six owners, who also began to acquire adjacent land once the route of the Beltway became apparent.[12] However, the nature of development at Tysons Corner and in much of Fairfax County is closely associated with the tenacity of the likes of John "Til" Hazel and Gerry Halpin, who, aside from having direct interests in development, also were instrumental in shaping and defending the county's aggressive pro-development stance since the mid-1970s.[13] Tysons is the product of the specific differential locational advantages partly set in motion by these state interventions.[14] Such investments ensured that "a new 'Americanism' even entered the language—'beltway'—to describe the broad expressways that encircled every important city by 1975 and that attracted employers of every description."[15] Tysons would be a perfect example of this phenomenon as it became the home to many "Beltway bandits," the moniker given to the many Pentagon defense contractors that congregated there from the 1960s on.

An interviewee recalled that the original growth entrepreneurs associated with Tysons Corner "were people who deal with maps and they don't deal with site plans. They look at regions and they say okay here are the pieces and where do you want to be in twenty years from now? I think they got this one right and they have been proven right by the success that Tysons Corner has had."[16] It is a description that resonates with Harvey Molotch's reference to the power of maps for the development community.[17]

After its beginnings in the late 1960s, "for the next twenty years, then, Tysons Corner, almost exactly halfway between the airport and the White House, was not only the first but also the last practical place for commercial activities between Dulles and the District since improvements to the limited-access Dulles Toll Road were not forthcoming."[18] It is a prime, although unusually concentrated, example of what Bruce Katz has called America's exit-ramp economy.[19]

Although the area developed along a quintessentially suburban road layout with suburban densities as a glorified suburban office park, the physical barriers presented by Routes 7 and 123, the Beltway, and the Dulles Toll Road, along with other planning decisions, such as that to maintain green and higher-density residential "buffers," also corralled development into a relatively compact triangle of land that is Tysons Corner, as illustrated in figure 6.2. The urban morphology of Tysons Corner consists of a jumble of retail and commercial buildings and associated activities (including a strip mall, auto dealerships, office blocks in their own seas of parking lots, and the retail megastructures of the two regional shopping malls and their associated parking), which are surrounded by highways and a sea of low-density suburban residential development in Fairfax County and the separate city of Vienna.

It was this triangular agglomeration, relatively compact though distinctly suburban in form, as it had emerged by the 1980s that Joel Garreau termed "edge city." The triangle of land that is formally designated the Tysons Corner activity area covers approximately 2,100 acres and straddles several magisterial districts within Fairfax County. It is not a single postal district, let alone a separate governmental jurisdiction.

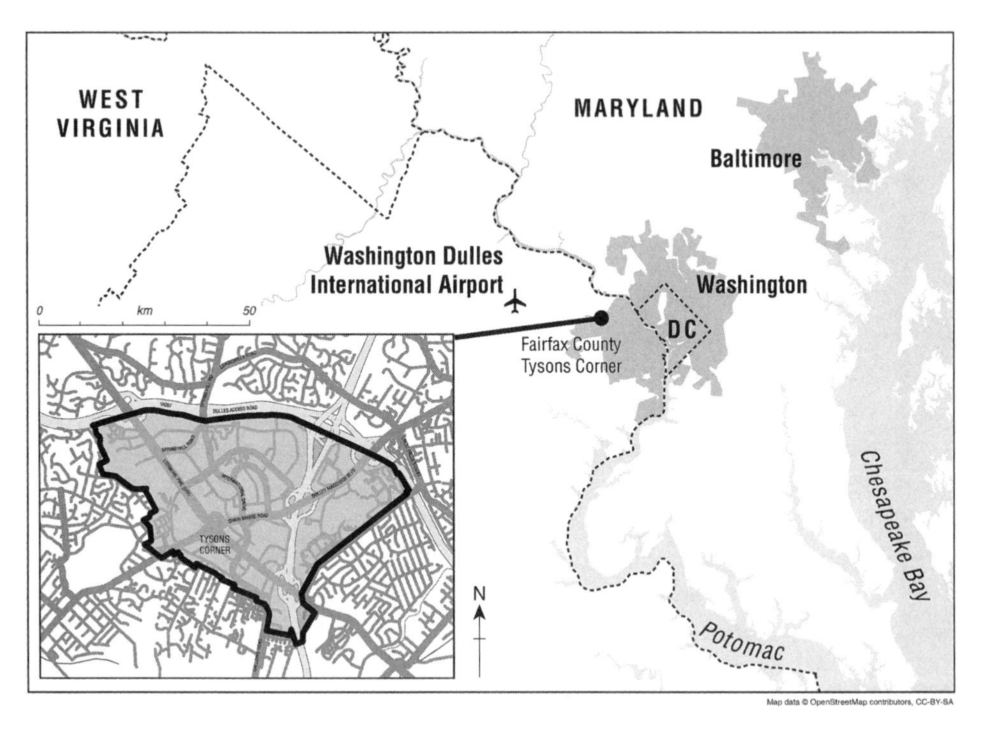

Figure 6.2
The Metropolitan Regional Context of Tysons Corner

Planning in the Face of Power: The Post-Suburban Politics of Self-Sufficiency

Tysons Corner also owes its existence to the sort of pragmatic post-suburban political bargain beginning to be struck by a number of suburban communities, including incorporated cities, but also counties across the United States, from as early as the 1950s and 1960s.[20] This bargain certainly did not arrive cleanly all in one piece but was one that gradually came to be accepted by the various interests—developers, politicians, and residents. It was a local post-suburban politics of pragmatism in the face of corporate power.

Much of the early history of the development of Fairfax County has been represented as one of a County Board of Supervisors often reluctant to accommodate population growth.[21] Nevertheless, Terry Spielman Peters was able to note how land-use planning policy in the county leading up to the first planning applications at Tysons Corner had in fact been led by the private sector (at that time preoccupied with residential development).[22] This was the case, then, with some of the first major industrial and commercial developments, such as the Tysons Corner Center, a shopping mall. As one interviewee recalled of the planning application for the Tysons Corner Center mall development in the late 1960s, "The first major project came along was what is now called Tysons 1 and they were so tickled to get the application for it that they waived all kinds of things. For instance, the road that circled it they allowed to be a private road. They waived the county parking requirements so that the whole thing started out with a deficiency of parking. They made no plans for any additional road improvements to carry the traffic to and from this major shopping center."[23]

During this time (the 1960s and 1970s), the zoning ordinance associated with the original 1958 Comprehensive County Plan was changed 144 times.[24] The county became a "textbook case of zoning irregularities and governmental mismanagement in the face of rampant growth."[25] In fact, county officials were jailed in the late 1960s following charges of corruption stemming from the early developments at Tysons.[26] Close on the heels of this scandal and the early developments at Tysons Corner came a short-lived antigrowth board of supervisors in the early 1970s.[27]

It was short-lived because economic crisis of the early to mid-1970s that signaled the end of the Fordist regime of accumulation ushered in what has been an almost consistently pro-growth board of supervisors, irrespective of overall political control of the county. Tysons Corner became the flagship of a political and popular pro-growth consensus in the county that can be traced to a special "blue ribbon" commission established at the request of the Fairfax County Board of Supervisors that examined the fiscal capacity of the county vis-à-vis perceived shortfalls in collective consumption expenditures.[28] In this respect, aggressive recruitment of businesses to Tysons Corner and later to areas such as Herndon and Reston, nearer Dulles Airport, by the Fairfax Economic Development Authority might be regarded as part of the planning of Tysons Corner conceived more broadly.

Moreover, in some sense this political bargain was one that could draw on an older, more established sense of acceptance. It has been claimed that as early as the 1950s and 1960s, the McLean Citizens Association had reached some sort of agreement with the Fairfax County Board of Supervisors that no commercial development would be allowed between Falls Church and McLean, effectively steering development toward the area of land that is now Tysons Corner.[29] Here, then, Paul E. Ceruzzi draws a contrast between the coalitions of interests that formed in Montgomery County, Maryland, and Fairfax County, Virginia, at the time the Capital Beltway was being constructed. In contrast to the already suburbanized Montgomery County, where established residents provided effective opposition to the routing of the Beltway, in rural Fairfax County, although some land had already been zoned for housing, developers delayed to fit in with the proposed route of the Beltway.[30]

The consistent pro-growth attitude of Fairfax County was one side of a political bargain that involved Tysons Corner funding the needs of a surrounding residential community. That is, growth machine politics in Fairfax County was nevertheless somewhat spatially restricted to or focused on Tysons Corner and later on the likes of Herndon, closer to the Washington Dulles International Airport. The conscious licensing of this growth machine grew out of a concern to relieve residents of some of the tax burden of funding collective consumption expenditures—primarily schools, for which Fairfax County has a national reputation—associated with the predicted population growth. However, it also had other aspects and indeed effects. On the positive side, and completing this particular post-suburban political "bargain" (as far as residents' participation in this suburban growth machine is concerned), was the fact that residential real estate taxes have been kept low in Fairfax County in comparison with neighboring Loudon County. One interviewee, alluding to the role Tysons Corner has played in allowing the county to fund such expenditures, and also to fund the development of infrastructure components such as water and sewerage systems, put it colorfully: "There is just no question that Tysons has been a tremendous benefit for the whole county. Is it the prettiest girl in the dance? No, but at least we had a pretty girl in the dance!"[31]

Nevertheless, on the negative side, interviewees cited a string of bitter battles during this period to win minor concessions from, and promises (subsequently broken) by, developers in Tysons Corner over the years.[32] In line with experiences internationally, planners and politicians have begun to bargain harder for developer contributions under the county's "proffer" system—a proffer system fashioned significantly, it should be noted, by those engaged in development and acting for the development industry.[33] These proffers have been on a case-by-case basis for each individual development proposal, with little thought expended on how they could contribute to an overarching pattern of development for the whole area.[34]

A Corner of Contradictions

As one interviewee from the business community in neighboring Vienna remarked of the growth of Tysons Corner to the present day, "It's reached critical mass, and now everybody is being forced to reclaim and make the best out of a situation that nobody paid any attention to to begin with. Everybody was just so happy that jobs were up, employment was up, buildings were going, taxes were good. You know, residents could have low taxes because businesses were footing the bill. Everybody was happy, happy, happy about all the economic development, and then all of a sudden while we were being happy this monster developed."[35]

Yet within this wider suburban fabric, Tysons Corner has continued to evolve and selectively deepen as a location for economic activity. As an interviewee argued, "When Joel Garreau called it an edge city it was an edge city. It was indeed a growing suburb. Tysons right now, in a business sense, is a city unto itself."[36] It is so much so that another interviewee highlighted the fuller ramifications of the growth of Tysons within the capital city region: "The District of Columbia is very threatened by us. . . . In private office space, out here in Fairfax County, we are comparable to DC. . . . You start wondering, are they our suburb or are we their suburb?"[37] This is no idle thought, as an editorial in the *Washington Post* covering the deliberations of the task force referred to Tysons Corner as the capital city region's second downtown.[38] Alongside developments across the metropolitan area, Tysons Corner has been a major element in "the transformation of Washington from a federal town to a major control point for advanced services."[39]

Tysons is a city in terms of its economic scale, but it exists in a suburban setting and with many of the trappings of suburbia. These trappings include an incredibly routine and even ugly set of commercial buildings for any potential downtown. The ugliness of Tysons Corner is the stuff of local legend, with some of its landmark buildings—including those by celebrated architects—referred to as "the toilet bowl," "the shopping bag," and "the pencil" (see figure 6.3). There are also important questions over the redundancy and commercial viability of some of the buildings found there. Some of these, such as a retail strip mall along Route 7 (visible as the

Figure 6.3
The "Toilet Bowl" Office Building at 2010 Chainbridge Road. *Source:* Author's photograph.

long building running horizontally in the first rendering in figure 6.5), are ripe for redevelopment.

In this context, the challenge that faces politicians, planners, the business community, and the residents of Fairfax County was noted by another interviewee: "Statistically it is already one of the top cities in the country, but when you think about the typical urban qualities of a city it contains next to none of them. There are no parks, there are no civic destinations. There isn't a mix of uses. There is an incredible imbalance between workers and residents. . . . Tysons isn't very livable . . . and it's not very walkable."[40] Here, then, some question whether the term edge city is an accurate one to apply to Tysons Corner. "Tysons is an overgrown suburb," said another interviewee. "The word city to me connotes a plan, or a direction or a civic quality, and Tysons has none of that."[41]

A Rage against the Machine? Growth versus the Environment at Tysons Corner

It is in this context that a Tysons Corner Land Use Task Force was convened in 2005 at the request of the Fairfax County Board of Supervisors.[42] Following the tradition for Fairfax County supervisors to defer planning matters to the single supervisor whose magisterial district Tysons largely falls into, the initial task force of thirteen members was relatively narrowly drawn. The Fairfax Chamber of Commerce and the original citizen members of the task force claimed a role in broadening membership to the eventual thirty-six persons to include other supervisors and business and citizen interests—although representatives from neighboring communities felt that such a seat at the table was rather hard-won.[43] Concerns remained that task force membership was skewed toward business interests, especially those with a vested interest in the increase in land and property values that would attend TOD densities.[44]

The task force explored a number of scenarios for the future development of Tysons Corner, predicated largely on the extension of the Washington, D.C., metro system through Tysons and out to Dulles Airport, as previous planning exercises had been.[45] The U.S. transportation secretary eventually approved federal funding as part of a tripartite package (federal, state, and county governments) for the metro extension to Dulles, and construction began, while a new comprehensive plan resulting from the work of the task force was adopted in 2010.[46] It should be noted that some elements of the plan to increase density, including residential development, have recently been put in place (see figure 6.4); under the new plan the density of development in Tysons Corner stands to increase markedly. The rendition of a before-and-after scenario for land adjacent to the metro line in figure 6.5 gives some indication of the low density of development that had existed even in a compact edge city like Tysons Corner, as well as the sort of transformation that might be expected. Put succinctly, the plan aims to transform Tysons from an edge city to a downtown for Fairfax County.

To this end, eight districts have been delimited, with the four centered on new metro stations being TOD districts. The plan calls for 75 percent of development to be within half a mile of metro stations, to permit a vertical expansion of Tysons into an urban center of 200,000 jobs and 100,000

Figure 6.4
New Condominiums Towering over Existing Residential Development in Tysons Corner. *Source:* Author's photograph.

residents. Indeed, Fairfax County approved the largest residential development in Tysons Corner in plans that envisaged an increase in an existing low-rise residential community, The Commons, from a mere 331 residential units to 2,571 units, with buildings rising to fifteen stories.[47] Notably, it is hoped that the residential development would also help stabilize or even create a better jobs balance of four jobs per household, and that development as a whole would finance a more environmentally sustainable Tysons featuring a network of green space and a redesigned transport system, including feeder and circulator connections to metro stations.[48] Development within a quarter mile of the metro stations will not be subject to a maximum FAR (floor-to-area ratio), with a recommended FAR of 2.0 for development beyond a quarter mile from stations.

Only recently has some degree of opposition to the further growth of Tysons Corner emerged, chiefly from the adjoining residential communities of Vienna and McLean. However, these seemingly "antigrowth" interests and any associated opposition to further development of Tysons Corner

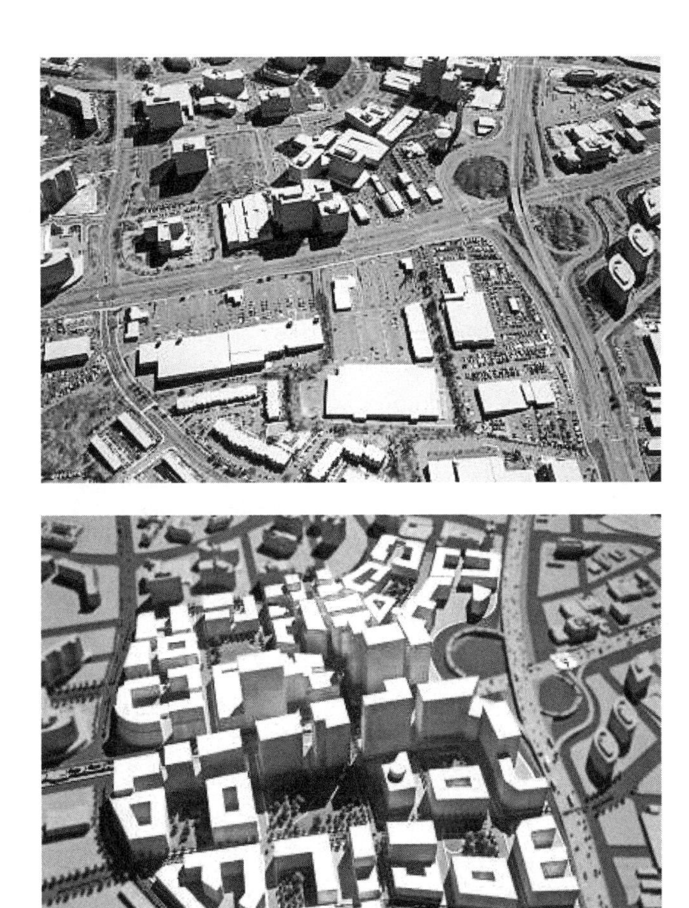

Figure 6.5
Rendition of Potential Increased Density of Development in Tysons Corner. *Source:* Reproduced with permission from Fairfax County Government.

can hardly be described as a rage against the suburban growth machine that has prevailed to date. Instead, concerns are more accurately characterized as having to do with the kind of growth being pursued rather than outright opposition to growth. In this sense, it is possible to see a selective deepening of the sorts of post-suburban sensibilities that had emerged quite early on in Fairfax and around Tysons Corner and its surrounding residential neighborhoods in respect to local self-sufficiency as the settlement moved from a suburban growth machine politics into a developmental regime.

However, there is little evidence as yet of the altered political inclination reflecting environmental issues or providing for collective consumption, which might necessitate governance arrangements of wider geographic latitude, such as at the metropolitan regional scale.

As an architect with a long-standing interest in Tysons Corner explained, 'The biggest battle that we have is actually to try to convince the county and the task force, and the citizens round about that higher density is not a four-letter word. Density is good. Density is healthy. Growth should be celebrated, not denied. It is going to grow anyway."[49] However, the Fairfax County Chamber of Commerce surveyed the population of Fairfax County and found a figure in the region of 80 percent in favor of metrorail to Tysons and approximately 60 percent in favor of TOD densities.[50] The McLean Citizens Association, for example, has an objective of supporting growth, and residents in the Vienna and McLean areas are doubtless aware of the benefits that a successful business community in their midst would mean in terms of lower taxation.[51]

There are a number of specific concerns regarding the impacts and amelioration of growth pressures that relate to the subject of this new compact. The main concern has to do with the additional traffic generated as a result of the increase in building densities over and above those buildings that will be removed as a result of the proposed extension of the mass transit line through Tysons Corner. The possibility of the metro system being extended through Tysons Corner has long been on the agenda. In this regard, it should be noted that as early as 1976, an informal planning exercise suggested that a metro line would not be of any substantial aid in alleviating the traffic problem at that time.[52]

A second set of concerns centers on preserving the environmental amenity enjoyed by Fairfax residents. A resolution from the McLean Citizens Association called on the Fairfax County Board of Supervisors, the County Planning Commission, and the Tysons Corner Land Use Task Force to "make the preservation of the quality of life in surrounding neighborhoods a top priority."[53] One specific concern has to do with rehabilitating Tysons Corner's own version of the Los Angeles River.[54] Several interviewees lamented the "loss" of Scott's Run, a creek that had been degraded and channeled through a series of culverts, with the consequent loss of wildlife habitat, as development progressed over the years at Tysons Corner.[55] Since "great cities are defined by great parks," current planning exercises for Tysons Corner

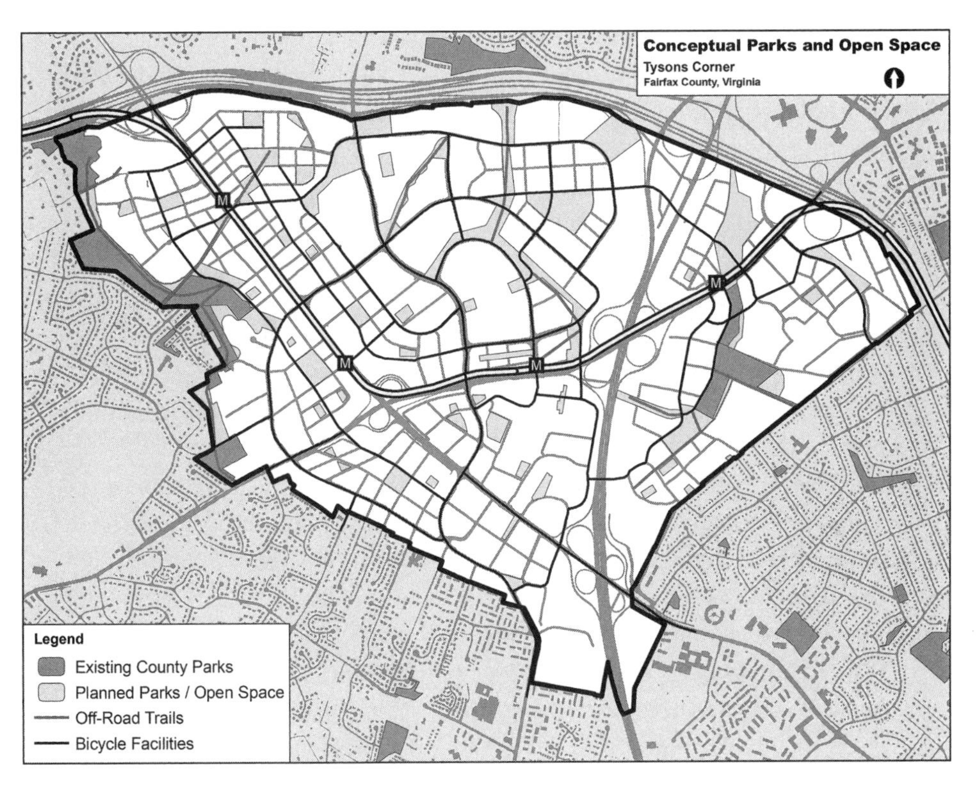

Figure 6.6
Proposed Network of Green Spaces for Tysons Corner. *Source:* Reproduced with permission from Fairfax County Government.

anticipate a network of parks and green spaces (see figure 6.6). As a draft nonregulatory planning exercise describes it, "As Tysons transforms from a suburban commercial center to a major regional urban center, a connected network of urban parks will help to distinguish Tysons as a great urban area and bring benefits to the local economy and quality of life."[56]

Though some "modernization" of Scott's Run Community Park has been given approval, it remains unclear quite what greater density of office and residential development will bring in the way of environmental conservation, let alone rehabilitation, since development and associated access proposals continue to place pressure on the remaining undeveloped parts of Tysons Corner.[57] Thus, suburban environmental politics centered on the future of Tysons barely touches on broader environmental concerns in a way that asks authorities to look beyond localized concerns and solutions.

While new developments in Tysons Corner will meet LEED (Leadership in Energy and Environmental Design) silver level certification to make a contribution to altering the carbon footprint of the place,[58] these improvements in the environmental performance of buildings feed into what was described earlier as technological solutions that fail to resolve the larger environmental impacts of the suburban way of life.[59] They also are evidence of something of a political and planning paradox in that those suburban jurisdictions that have made mistakes in the environmental and planning fields now have the revenues and the support of the population to engage in conservation measures, whereas rural jurisdictions in the process of becoming suburban lack both the revenue streams and support from the population to plan effectively for conservation.[60]

However, these developments in building standards, and as a response to regional, national, or even global environmental issues such as the effects on water catchments, land take, and global climate change, hardly embody the sort of global sense of place that would drive collective responses to the sorts of carbon footprint of suburbanization as a whole at the Washington, D.C., metropolitan regional scale, let alone beyond. Instead, specific inter-jurisdictional concerns center on spillovers in the density of development being sought at Tysons Corner in the communities of McLean and Pimmit in Fairfax County and in the separate city of Vienna. There is also some awareness among the residents of the resentment felt elsewhere in the state over money meant for alleviating traffic congestion being spent specifically on Tysons Corner.[61]

Instead, the environmental concerns of residents in Tysons Corner continue to revolve around traditional concerns of residential and local environmental amenity. Thus, concerns continue to invoke local self-sufficiency by attending to the adequate provision of amenities such as school playing fields, parks, and sports facilities, as well as basic infrastructural components such as water supply and sewerage systems.[62] The regeneration of the nearby Ballston-Rosslyn suburban corridor is the local model drawn on in current planning deliberations.[63] However, in contrast to the Ballston-Rosslyn corridor, the extension of the metro system through Tysons Corner will be elevated, and the "it's not over until it's under" campaign has focused on the need for the extension of a metro line to Tysons Corner to be underground if the suburban, auto-dependent format of Tysons is to be improved significantly.[64] Factored into plans is therefore a substantial increase of 154 acres of new parks and public spaces, in contrast to the almost complete absence of public space currently in Tysons Corner.[65] The very localized focus of environmental concerns seems likely to be underlined in the immediate future by a growing residential population in Tysons Corner itself, which may go far toward ensuring that developers have fewer opportunities to renege on commitments to parks and open spaces, as was the charge with developments of the past. However, quite how population increases—including a provision for 20 percent affordable housing in developments—will dilute the limited liability of local politics in Tysons Corner and Fairfax County in the longer term remains unclear.

Perhaps as a result of the limited liability of Fairfax residents and county government in the face of the power of private interests in these regards, many question whether the sum of these developments (including the overarching vision of the Tysons Corner Land Use Task Force) amounts to a real departure from existing auto-oriented suburban formats of development.[66] In a passage that resonates with Jon Teaford's depictions of a distinctive post-suburban ideology and politics, an interviewee from the Fairfax County Chamber of Commerce said, "We are not going to replace the suburb. What we are talking about is creating some urban centers that will provide the capacity to manage our growth. . . . People in this country identify with their neighborhood school, with the local place for worship, their neighborhood's swimming pool, or the homeowners' associations. . . . We are not going to disrupt that core sense of identity, but we are going to take those places where we can put the mass transit, we are going to redevelop those with an intensity that is much different than now and in doing so we are going to protect and preserve what is been built over the last 50 years."[67] Indeed, it has been suggested that for their part in the work of the Tysons Corner Land Use Task Force, county planners had been "approaching the redevelopment of Tysons with a suburban mentality, reluctant to embrace the kind of urban place Tysons needed to be."[68] Thus, as a result of residents' sensibilities, developer sensitivities, and even county planners', Tysons as a post-suburban moment is likely to be a hybrid rather than something entirely different.

As I discuss in passing in the next section, as a hybrid accommodation between traditional residential suburbia and urban density, Tysons Corner will also coexist with developments of suburban residential and commercial development density and format that represent business as usual farther out in the restless metropolitan urban regional landscape.

Turning the Corner? Collective Consumption Needs and the End of a Suburban Growth Machine

As one might expect in a now sizable and mature edge city, some of the diversity of Tysons Corner and the legacy of its past growth linger, in the sense of there being different business and developer interests with different time horizons.[69] The greatest contrast is between the short-term suburban growth machine interests that historically have driven the growth of Tysons Corner and the longer-term interests of landowners and building occupants who, while also part of this suburban growth machine, have aligned themselves with the agenda anchored in TOD, or "smart growth." In sum, then, urban politics in Tysons Corner might be portrayed as standing at the forefront of what I depicted in chapter 3 (figure 3.1) as the growth machine associated with new suburbs and a developmental regime associated with the evolution of suburbs into post-suburbs. The salient point here is that state interventions stand to have played a decisive but contrasting role in the historical and future development of Tysons Corner, the area's initial growth having been shaped significantly by investment in roads for the private automobile and associated suburban land-use format and its future development (or retrofitting) crucially dependent on significant state investment in public mass transit and an urban format of development.

One group of developers has pro-growth agendas that might conveniently be labeled the suburban growth machine business as usual. The interests of this group are necessarily much wider than the already built-out Fairfax County and Tysons Corner areas. Some remain attached to the way Tysons Corner is and have already turned their attention instead to development opportunities around what was envisaged as an "outer beltway" in the 1960s but was never built except by increments in the form of county parkways. For example, Til Hazel, one of the original and most tenacious growth entrepreneurs of Tysons Corner, had in the intervening years attempted to replicate a Tysons Corner–style development on land adjacent to the Manassas Battlefield National Park, provoking a national outcry and the expensive repurchase of land by the federal government.[70] In this instance, Prince William County supervisors had been anxious to broaden the county's tax base and stand out from the shadow of neighboring Fairfax County in economic development terms.[71] Not surprisingly,

Hazel remains proud of the edge city Tysons Corner is. While apparently amused at the recent unofficial dropping of "Corner" from Tysons Corner in the work of the Tysons Partnership, he nevertheless is clear that "it's the most successful edge city in America. . . . Why do you want to keep messing with it?"[72]

Relatedly, one interviewee noted that in contrast to his organization's preference for TOD at Tysons Corner, he was "convinced that there is one type of development out there in particular that also has a model of buy land cheap, especially farmland, then work the political system to get the comprehensive plans and zoning changes, then work the system to get the infrastructure investments."[73]

In their specific concerns regarding the current deliberations of the Tysons Corner Land Use Task Force, this group of development interests has unwitting allies in the real estate brokerage industry, which has tended to advocate competitor locations to Tysons. Since the 1980s, Tysons Corner has become part of a much more extensive, forty-mile-long Dulles Airport corridor of business activity.[74] The development industry and the real estate brokerage industry have interests that spread across this larger corridor, in which there are qualitatively different business propositions. As an architect highlighted regarding the differences between Tysons Corner and the planned community of Reston, "They are only about five or six miles apart, but it's worlds apart in terms of how they position themselves and how they market themselves and how the market sees them. The rents in Reston are now more than those in Tysons. That wasn't true even ten years ago. . . . People are voting with their wallets for what I would call a sane, planned, urban design and voting in a certain way against the unorderly sprawl of Tysons Corner."[75] Specific developers and real estate brokers have also tended to highlight the time-limited costs of the disruption that is likely with the arrival of a metro line and the numerous associated building constructions. To a limited extent, these suburban growth machine business-as-usual interests have also been keen to relay resident groups' skepticism of the benefits of increased density and their desires to preserve Fairfax County's suburban nature.[76]

The debates and contests are ongoing, although it would appear that for now, those interests implicated in the suburban growth machine that drove development at Tysons Corner in the past are no longer ascendant. Instead, real estate and developer interests have aligned themselves strongly with the notion of fashioning a city from Tysons Corner and have been prominent voices in the direction taken by the Tysons Corner Land Use Task Force. This group of developer interests includes some of the larger and, in most cases, original landowners, who, it must be said, see the economic benefits of the increase in land values and rents that would attend a greater density of development in Tysons Corner. However, supporters also include aggregations of some of the smaller businesses operating in Tysons Corner.

The interests of these developers are driven by some of the new realities faced by developers at Tysons Corner simply as a result of its maturing as an already developed location. The challenges mirror those faced by suburban communities as whole, as indicated in the discussion in chapter 4 (see also figure 4.1). At the outset of its growth, the land of Tysons Corner was owned by relatively few people. Today the chair of the Tysons Corner Land Use Task Force in the region estimates there are 150 landowners.[77] In some respects the growing complexity of the redevelopment challenge in Tysons Corner belies the apparently simple and rather random scattering of its suburban office and retail appearance, for Tysons Corner has been subject to at least three distinct bursts of property development, with a few individual sites having been developed or redeveloped during each round.[78]

Of perhaps greatest significance and immediacy regarding prospects for the transformation of Tysons Corner are the individual corporate master planning projects of some of the largest and longest-standing real estate interests in Tysons, which seek to embrace the TOD aspirations of the task force. Macerich, which owns and operates the original Tysons Corner Center shopping mall, has plans that antedate the current task force deliberations and received zoning approval under the existing land-use plan for Phases 1 and 2 of a four-phase redevelopment of its site.[79] Most notably, West Group, still the largest single landowner in Tysons Corner, has its own plans for a greater density of development based on the arrival of a metro stop. West Group has gradually divested some of its land and buildings to former tenants; however, its plans for the vast area that it retains in Tysons Corner are striking. As one interviewee said, "Our partners . . . asked us to come up with a new long-range plan where every one of our twenty-five buildings comes down. . . . We have been in Fairfax County for fifty years, we intend to be here for the long haul. . . . We have taken the task of stewardship [in] coming up with a new vision for the future very seriously. . . . We think the opportunity in front of Tysons right now is equal or greater than when they bought dairy farms because two roads were crossing over it."[80] Here, as one interviewee explained, the sort of redevelopment envisaged in Tysons Corner becoming more urban will nevertheless be hybrid: "The effort to make Tysons more urban is a challenge because Fairfax County is still very suburban. So you will notice things that you wouldn't see in a city, like these bizarre, large paved areas that are really fire lanes. . . . You get some strange little suburban and urban influences that meld together on these projects because you are trying to create an urban environment but you are working with the constraints of infrastructure that is truly suburban."[81]

Echoing the sentiments voiced above, another interviewee went on to stress the irreversibility and long-term view that would necessarily be associated with public TOD densities.[82] For one interviewee, the historical and future development of Tysons Corner stood in contrast as two distinct models of development: "We realized that we can't continue to develop Tysons in the same way it was developed in the past. Now, that is not to say that it was done wrong in the past. . . . Tysons was initially developed as a suburban office campus; it wasn't developed to be a large urban area. So when we thought about how does it need to grow going forward, we realized we needed a new development model."[83]

Beyond these individual business interests, there have been two notable attempts to bring together groups of smaller landowners to facilitate the reworking of plots of land and road layouts in Tysons Corner. The most successful of these attempts was the demonstration project of the Georgelas Group, which entailed the consolidation of the group's own twenty-eight acres of land and properties at the western edge of Tysons Corner.[84] Davis Carter Scott Architects and Planners worked with around thirty different landowners (primarily auto dealers) to rationalize land parcels, create a grid of streets, and release land for development, though this effort subsequently appeared to run into difficulties owing to the number of parties involved. The possibility of fashioning a grid of streets within Tysons Corner was raised in one early thought piece, "Taming Tysons," though these two trial initiatives also reveal the scale and degree of difficulty of the task at hand in this respect.[85]

A City in Waiting? Government or Governance for Tysons Corner

One of the ironies of land-use planning in America is that it is the capital city metropolitan area that remains perhaps the least well served by its metropolitan governance arrangements. The difficulties of constructing regional governance arrangements doubtless stem from the tristate nature of the functional economic scope of the metropolitan region. From its earliest days, Tysons Corner therefore grew in something of a vacuum of county government and planning within a larger vacuum of metropolitan regional planning and government. And although it is the county scale that Jon Teaford sees as the primary scale at which America's post-suburbia has been sutured together, it is clear that some involved with Tysons today consider that the county itself has not evolved sufficiently from its rural roots: "Fairfax County should have evolved away from a county a long time ago because it is not like any foreseen county . . . we have all the attributes of a city."[86]

While county approaches to planning have come a long way since the 1960s, the same cannot be said for the influence of metropolitan governance arrangements. These are organized under a Metropolitan Washington Council of Governments (MWCOG) and supplemented by the coordination offered by nonprofit organizations such as The Greater Washington Initiative focused on economic development. The MWCOG's regional plan, *Region Forward,* covers many of the staples found in other metropolitan regional planning documents across America. Following earlier identification of and focus on regional activity centers as an appropriate planning tool to address many of the region's land-use, transportation, and environmental issues, the current plan seeks to concentrate future development, transit accessibility, and a mix of housing in the regional activity centers that currently account for the majority of employment across the region.[87] However, MWCOG arguably enjoys very little in the way of leadership in advocacy or power to orchestrate coordination over investment and environmental issues. One indication of this is the newfound urgency in the creation of the Greater Washington 2050 Coalition in 2008 and the fact that the momentum to create a comprehensive regional vision has only been growing since the millennium, with "frustration that the 'business as usual' approach to these challenges would limit future success."[88] This is true even for an interviewee from an organization lobbying for greater road development and accessibility on the metropolitan region. "The nature of the Metropolitan Washington Council of Governments [to be] dominated by local governments also makes it difficult. The elected people who serve on the Metropolitan Washington Council of Governments have a very parochial outlook. Their first job is to lay defense. . . . The next thing is that you try to find the lowest common denominator, but that seldom rises to the point of what is the big picture, and they won't deal with that."[89]

At the local scale of the county, it could be said that the growth of Tysons was channeled and contained not only by the physical barriers of the main roadways but also by the voters of nearby residential areas. The main debate now exists over the precise format by which the greater densities of development at Tysons can be achieved. The major concerns regard the ability of any chosen "delivery vehicle" to implement the new land-use plan and to police the provision of infrastructure concurrent with development.[90] This is the new bargain among county government, developers, and citizens regarding the future development of Tysons Corner.

Despite claims for Tysons as "America's next great city," there is almost zero prospect of Tysons Corner becoming a separate governmental jurisdiction, given its fiscal contribution to Fairfax County.[91] As one interviewee elaborated, "The reason I believe that it would never become a city is that there is no incentive for the county to let that entity—which is the cash cow—get away. Because of all the businesses and all those millions of dollars of real estate taxes are coming into the county, they will never let it secede. But that doesn't mean that there isn't hope for a more civic Tysons."[92] If there is not to be a separate government for Tysons Corner anytime soon, the question remains whether this hope can be realized under the governance arrangements that have been put in place.

Instead, as in other edge cities in the United States, much of Tysons Corner has grown and has been administered under the auspices of the several "shadow" or private governments of developers, property owners and operators, and gated residential communities (of which there are several) (see figure 6.7).[93] In this respect, there is an intriguing open question regarding whether the private residential and corporate power interests ranged against efforts to plan Tysons Corner may yet prove a force for good in the reworking of suburbia in this instance. As one interviewee noted, "We built our road system around a one family car. . . . There are now many more car stops . . . groceries, laundry, kids, in separate locations. The cost

Figure 6.7
The Rotonda Gated Residential Community. *Source:* Author's photograph.

of putting the genie back in the bottle is huge, and people don't want to do it."[94] If suburban residents are themselves reluctant to foot the bill for things like mass transit extensions and developments through sales and property tax levies, then in some instances the business community that moved en masse with them may be willing and able. The business community itself has lobbied for the creation of special taxation districts in Tysons Corner and Reston to partially fund the extension of the metro system to Washington Dulles International Airport.[95] On the one hand, the business community's intervention appears likely to have been decisive in that the taxation district is effectively paying the state's share because of the historical disinterest of the state senate and legislature, located in Richmond, in Fairfax affairs. Despite much business having by now moved to the suburbs, and despite suburbia being far more commercially based than is often appreciated, business politics and business involvement in wider government agendas are far more immature than one might suspect. There may be relatively few places in the United States that can call on this sort of business involvement and funding for projects of collective consumption, even in a suburban economic powerhouse such as Schaumburg, Illinois, as we shall see in the next chapter.

On the other hand, in light of the historical difficulties of obtaining developer contributions (or proffers) in Tysons Corner, there remain suspicions that the civic space and amenities being sought will not materialize because of the reluctance of individual developers to pay for these. One view was that in the absence of significant funding from the Virginia Department of Transportation for the metro line extension, the private sector had already paid for much of the existing infrastructure in Tysons Corner, and a delicate balancing act was warranted regarding further demands on the private sector going forward.[96]

The new comprehensive plan for Tysons adopted in 2010 had twenty follow-on motions relating to its implementation covering infrastructure funding, the Tysons Partnership, legislative items, plan monitoring, and transportation and facilities planning. The "Tysons Partnership was established by the Fairfax County Board of Supervisors to help implement the overarching values and goals embodied in the Comprehensive Plan for Tysons" so that supplemental guidance can be given on issues that arise in the course of development and that are not already addressed in the plan. Advice is taken from six Partnership Councils that facilitate stakeholder dialogue covering facilities and amenities, finance, marketing and branding, sustainability and stewardship, transportation, and urban design.[97] Here important questions arise over the evolving relationship between these shadow governments and the new governance entity created under the auspices of the county government.

It is tempting to think that a brash new settlement such as Tysons Corner has sprung up spontaneously in the liberal market context that is the United States. However, Fairfax County plans have been implicated in the format of Tysons Corner development, and specific plans for Tysons Corner have existed since the 1960s. It is curious that the "monster" that is Tysons Corner stands just a few miles away from, and in some contrast to, the celebrated planned community of Reston, conceived by Robert Simon around the same time that Tysons Corner first began to grow. Both have had their troubled histories in planning terms, while, as an interviewee noted, some semblance of Reston's planned coherence is only just beginning to become apparent.[98] This gives some indication of the time over which planning visions, however, tightly controlled, can take to come to fruition. As the same interviewee described, "Tysons, we keep planning it, and some day we hope it takes and we get it right. It is always on the edge of becoming." While there are important question marks over the possibilities of transcending its suburban origins, the clearest signal of such a likely change in the dominant mode of urban politics shaping Tysons Corner would be for an overgrown suburban business park favored by past state interventions in roads for the private automobile to be favored in the future by state interventions in public mass transit and all that that entails in terms of building densities.

Conclusion

The "separatist geography" of office campuses may well pose a major "problem for the future reform of the suburban landscape—the expectation that the process of suburban densification can ultimately be a solution to an unsustainable metropolitan form."[99] As a slightly more permeable and diverse collection of corporate office developments and mini-campuses, Tysons Corner may prove to be more reworkable than the typical singular suburban office campus that Mozingo describes. However, of a number of suburban development formats, edge cities were considered to pose the most daunting of retrofit challenges in one recent study.[100]

While the Tysons case is perhaps unique in terms of the major investment in public mass transit that is being made and that will facilitate a greater density of development and reworking of street patterns, it is also indicative of how attitudes among the private sector may yet evolve to offer up possibilities for a more urban, transit-oriented future for suburban communities. Nevertheless, the remaking of Tysons has a much wider significance, as the chapter's epigraph, a quotation from an architect involved in an early demonstration project concerning the potential for "taming Tysons," suggested.

The development of Tysons Corner into a more balanced post-suburban community with greater densities of employment and residential land uses but also potentially the retrofit of urban functions such as civic space and amenities implies a different model of development. It also appears to imply, indeed to demand, a different urban politics, and a maturation and diversification of the landed business interests associated with that politics. Of course, there is enough in the history of Tysons Corner to question the depth of some of these future visions, notably the aspirations for a more pedestrian-friendly business center with civic qualities. Nevertheless, a number of new development schemes have been approved that project a significant increase in the building density and residential population of Tysons Corner.[101] Already some evidence of a Tysons version 2.0 is emerging.[102]

Ultimately, though, any conclusions regarding this particular post-suburban moment in the history of the American outer suburbs must remain provisional. As one newspaper report has it, "Can an American city—a place of authenticity, culture and even grit—evolve out of a 50-year-old place built with none of that in mind? More than 20 years into the planning, it is unclear."[103] More fundamentally, this present-day test case for the reworking of America's suburban expanses may also already represent something of a backward glance, since Tysons Corner forms part of a much larger "edgeless city" string of employment and residential centers that includes Herndon and Reston. If, as Robert E. Lang suggests, American suburban development only briefly touched down in the relatively compact form of edge cities such as Tysons Corner, then we need to look elsewhere to find a more faithful account of any sequel to suburbia.[104]

7 Schaumburg: The Post-Suburban Future Yet to Come?

This concept calls for high-density living and working. . . . The environmental and open spaces of the region can be more readily preserved when the population is concentrated in smaller areas. . . . The outer planets is an original contribution to the philosophy of master planning—and may well set the trend for future urban and suburban design.

—Schaumburg Outer Planets Corporation, *The Outer Planets: A Regional Master Concept Planned Unit Development*[1]

The epigraph to this chapter is taken from a prospectus for a speculative high-rise New Town development. What was being envisaged for the young incorporated Illinois community of Schaumburg, composed largely still of open ground and detached single-family houses twenty-six miles from downtown Chicago, was remarkable. Just as remarkable was that the proposal was granted planning permission. Had it been built, it would have resulted in a very different Schaumburg from the city that exists today. However, in other respects, its rhetoric—of the discovery of the rapidly emerging outer suburban planets of Chicago—is as accurate today as it was then; it is just that this vision has been realized in horizontal rather than vertical terms.

In one important respect, namely, the early attempt to balance suburban residential sensibilities with the fiscal realities of self-government, the Village of Schaumburg was "born" post-suburban. It was born with, and its politicians and residents retain a vision of, local self-sufficiency, or what I have termed a mark I, locally oriented, post-suburban politics. Incorporated with a population of fewer than two hundred people in the 1950s, Schaumburg grew rapidly in land area as a result of aggressive annexation and was conceived and planned almost from the outset as a new kind of city, a regional capital for the northwestern suburbs of Chicagoland. Yet its conception as a particular and new type of very diffuse suburban city also in some ways has entrapped it more fully than suburbs such as those of Fairfax County and Kendall, considered in the preceding chapters, which show some signs of evolving into post-suburbs. Schaumburg provides a glimpse of the difficulties of building post-suburban communities from the majority of what Robert E. Lang has described as the "edgeless city" suburban expanses of America.[2] The extreme separation of land uses, the very low density of development, and limited possibilities for extensions of public transit, even in the older, public-transit-rich metropolitan context of Chicagoland, speak powerfully to the size of the challenge of building post-suburban communities in America. Suburban contradictions endure and are likely to magnify in post-suburban Schaumburg. The limited liability of the community's local self-sufficiency cannot obscure the regional scale of the task of fashioning a mark II post-suburban politics involving inter-jurisdictional cooperation with respect to both the balancing of the pursuit of private accumulation and conservation of the environment and making some provision for collective consumption needs.

Schaumburg has benefited from remarkable continuity and stability in political leadership since its incorporation, though important questions remain as to how political leaders will be able to engage and take the resident population with them as they continue to shape this expansive and new kind of outer city in function but not in form. While separate status as an incorporated community may have been part of the historical solution to the limits of suburban communities, it is not clear whether it will be enough for a post-suburban future.

Paternalistic Politics and the Building of a New Kind of Suburban City

"At the turn of the 20th century, Schaumburg was a community largely isolated from the mainstream of American culture by language, geography, and economics," the historian William J. Holderfield writes.[3] This situation was not to change until the 1950s, when Arizona residential builder Jack Hoffman's company F&S (Father and Son) acquired unincorporated land in Cook County, Illinois, and began building a suburban residential tract known as Hoffman Estates. "Hoffman felt that that the suburbs that would carry his name signified the pinnacle of his career as a home builder,"

Cheryl Lemus observes. "This was his Levittown. . . . But what Hoffman envisaged soon butted up against residents' questions of his paternalism. For the first five years, issues regarding incorporation and changing a namesake divided this burgeoning suburb."[4]

Almost immediately, in 1956, the nearby small community of farmers, descended from German immigrants in the 1700s, moved to incorporate as the Village of Schaumburg.[5] The village incorporated itself with just 130 farmers as a population—certainly not enough to support the running costs of the school district associated with the new community. Among the initial population was a sprinkling of wealthy "gentleman farmers." One of them was Robert O. Atcher, a "singing cowboy" entertainer of some note in that era but someone who also had a vision of a new kind of community, a new kind of city, that could be fashioned in the still largely undeveloped land to the northwest of Chicago. While the paternalism of Jack Hoffman proved a sticking point for an established if very new population, the paternalism of Robert Atcher became the driving force behind the community of Schaumburg.

Although F&S were despised by the farmers of Schaumburg, their arrival and their breaking of ground for Hoffman Estates suggested to Atcher that development was coming, and that it would be best to plan for it comprehensively rather than accept it piece by residential piece. Thus it was that what began as a reactionary response to suburban housing tract development quickly evolved into a forward-looking vision that, as a first step, saw Schaumburg embark on an aggressive program of annexing land to expand from 2 km^2 to 19 square miles. The curious boundaries of Schaumburg (see figure 7.1) reflect how, in the process of annexation, it virtually encircled its neighboring community Hoffman Estates during the late 1950s and early 1960s, in the process igniting a bitter rivalry between the two, as the latter managed to incorporate only some years later. Schaumburg limited Hoffman Estates' possibilities of expansion to a narrow corridor to the west, and also frustrated Hoffman Estates' attempts to become a village.[6]

To some extent, the Chicagoland region had a muted polycentric framework from early on, with important railway towns such as Joliet, Elgin, and Waukegan having grown historically in tandem with Chicago and now again experiencing growth, including transit-oriented development (TOD).[7] To the northwest of Chicago, major railway centers included Arlington Heights and Barrington. In contrast, the part of Cook County

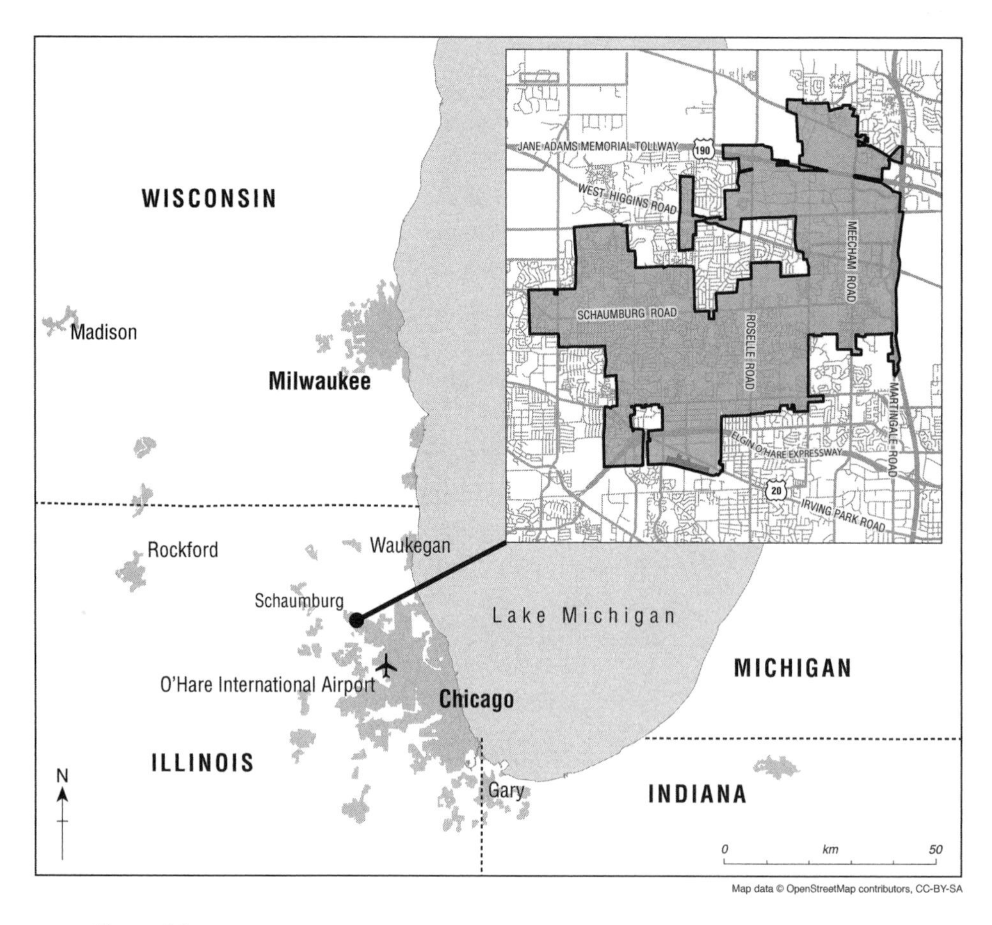

Figure 7.1
The Village of Schaumburg in Metropolitan Regional Context

where Schaumburg and Hoffman Estates are situated was poorly served by rail. However, Schaumburg emerged as a new and different point of growth in this polycentric suburban pattern, basing its development on the car and the airplane, since by the 1950s federal and state governments had approved the building of Interstate I-90 and the conversion of O'Hare military airfield into a commercial airport. As Schaumburg's current mayor, Al Larson, summarized: "Most towns around here developed around train stations. We are an exception to that. We grew up around the automobile. We knew development was coming out here. . . . It was going to happen whether you like it or not, so how best can we channel that development to benefit the residents and the village. And we deliberately had that regional center over here to keep that regional traffic away from the residential neighborhoods."[8] Moreover, the early vision was of a community that would be self-sustaining in tax terms. From the very beginning, Mayor Atcher had a vision of a complete urban community—but one built ostensibly at suburban densities and with strict separation of land uses. The idea of a growing, self-sustaining community is affirmed in the Village of Schaumburg's written profile of itself, which includes the following reflection: "As the web of urbanization spread to the northwest suburbs, the village was prepared, not just to accept growth, but to plan for it in an orderly fashion. . . . From the start, Schaumburg was planned not to be just another bedroom community, but rather the hub of commerce and industry for the northwest suburbs."[9] Exemplifying the marriage of suburban residential ideals and economic realities that Jon Teaford describes on observing the struggle of American suburbs to become post-suburban, Schaumburg leapt over that effort by being *born* post-suburban. However, in form, it was perhaps the ultimate logical expression of a modern, automobile-oriented suburb and associated land-use separation.

Paternalistic Politics and Schaumburg's Mark I Post-Suburban Fix

The rash of suburban residential developments popping up across America were also, significantly, ones that were mired in issues of political and bureaucratic corruption. They were growing for the most part in rural counties or as newly incorporated suburban communities whose immature and as yet poorly resourced governments and their political representatives were exposed for the first time to serious and regular dilemmas of a personal and community-wide scope regarding zoning applications, often in the absence of an adopted comprehensive land-use plan and accompanying zoning ordinances. As we saw in the previous chapter, the conviction of Fairfax County officials provided part of the context for understanding developments at Tysons Corner. The jailing of elected officials in Hoffman Estates, Schaumburg's neighbor, in the early years following incorporation also provides a context for, and contrast to, politics and development in Schaumburg.

Although politics in Schaumburg has definitely been of a Republican hue, local politics has been pursued outside the Republican Party machine. Almost all of the history of the Village of Schaumburg coincides with the leadership provided by two long-serving mayors, Robert Atcher (1961–1976) and Al Larson (1976–present). Despite some differences of view between the two, under the leadership of both it proves rather difficult to characterize Schaumburg politics in simple left or right terms. While the fiscal base was conceived to appeal to staple Republican political desires for low taxation by obviating local property taxes, the political and bureaucratic leadership also has included instances where the increase in land values accompanying zoning has been extracted in the form of developer contributions, to be recycled into community facilities and services, including green space and Schaumburg's own civic center.

The people of suburban communities and their governments are acutely aware of social distinctions, as in references to "Rolling Ghettos" for Rolling Meadows, which neighbors Schaumburg. And yet, while no one could accuse Schaumburg of providing very fully for affordable or social housing and a mix of social classes even in comparison with its near suburban neighbors such as Hoffman Estates and Rolling Meadows, there are—perhaps uniquely even among wealthy American suburban communities—no gated residential communities.

Finally, by the 1980s, Schaumburg had already emerged as a suburban employment node of some significance within the metropolitan region.[10] As the village's own material describes it: "The village of Schaumburg is the quintessential edge city. . . . But Schaumburg offers much more than just the stereotypical edge city. Since its incorporation in 1956, the Village leaders have strived . . . to create a complete community that is able to focus on the residents and their needs as well as the businesses and their needs."[11] Opinions differ on the extent to which Schaumburg has become self-sufficient in terms of its economy, with interviewees stressing the mix of businesses represented in the village and its having become something of a destination.[12] Others note that, despite the size of the local office market in and around Schaumburg, the sense is still that Schaumburg relies on the growth spilling out from Chicago.[13] The process of rounding out the citylike character of this essentially suburban community continues with the recent creation of a convention center: "The mayor's vision of the convention center was to facilitate that. So businesses do not have to go downtown . . . they can hold events, conventions, conferences at a facility nearby. . . . Corporate groups love Schaumburg because everything comes together in one place."[14] Mayor Al Larson is also very conversant with current academic and policy discussion surrounding Richard Florida's notion of a "creative class" and its contribution to local economic development. Yet with all that these ideas imply regarding the recruitment of industry, commerce, and talent, again somewhat distinctively compared to other neighboring communities in the northwestern suburbs of Chicagoland, the Village of Schaumburg also refuses to offer any incentives when recruiting industrial and service sector employers.

The aggressive annexation of county land was at least partly intended to provide for the large expanse that would serve as the tax-generating commercial heart of newly formed Schaumburg. From the very beginning, Atcher envisioned Schaumburg as a regional commercial center, that is, as a commercial center for the northwestern suburbs region of Chicago. As the second mayor, Atcher, together with long-serving attorney Jack Siegel, drew up a zoning ordinance for a population yet to arrive and for land largely empty at the time. This early zoning ordinance provided for a large area of land to be reserved for a regional commercial center even though there was as yet no market for commercial space at this distance from Chicago.[15]

In parallel with events in Tysons Corner and Kendall Downtown, recounted in the preceding chapters, developers in the area began to take a real interest in Schaumburg's regional commercial center in the late 1960s and early 1970s. Interest began with the attraction of corporate headquarters campuses of the Pure Oil Company and Motorola in the 1960s, which, like office parks elsewhere in America, formed a distinctive low-density landscaped part of the emerging suburban expanses of America. However, perhaps the biggest single element in the suburban matrix of Schaumburg, and one that continues to underpin the fiscal health of the community, was Woodfield Mall, a shopping complex that opened in 1971. Eventually the mall became surrounded by office developments as the likes of insurance companies moved some of their operations out of Chicago.

If takeup of the land earmarked for commercial uses was slow, one small irony was that the new Village of Schaumburg was not slow in approving residential development of a character little different from that which had caused such consternation over the F&S developments in neighboring Hoffman Estates. Alfred Campanelli, from New England, began building detached Colonial-style suburban residences in the southwest of Schaumburg (see figure 7.2). In Hoffman Estates, F&S had provided schools and

Figure 7.2
An Alfred Campanelli Residential Development in Schaumburg. *Source:* Author's photograph.

other facilities to accompany the new residential community that was emerging. In Schaumburg, the village bargained so hard with the developer to extract contributions that at one point the developments were referred to as "Campanelli's graveyard."

Schaumburg's Vertical Future That Wasn't

Suburbia as we know it also, in perhaps countless cases untold, contains the seeds of alternatives left ungerminated—some radically different from what we have come to know. One such is the Schaumburg Outer Planets Corporation's master plan for Schaumburg. It was a vision of high-rise developments very different from the Schaumburg of today, and yet one that received planning approval in the late 1960s.

As described above, Mayor Atcher sought to populate the regional commercial center and other parts of Schaumburg with developments of a regional stature. He was also persuaded of the merits of planned unit developments (PUDs), which were self-contained in terms of their utilities and infrastructure and made few current demands on the municipality with respect to maintenance of infrastructure. Several such small-scale residential PUDs had been built in Schaumburg by the 1960s. Onto this scene came the quite amazing PUD scheme of the Schaumburg Outer Planets Corporation, led by local developer Lee Romano, for a highly accessible parcel of land close to the intersection of the I-90 and I-290 expressways, which received planning approval in 1968. The scale of the planned development as outlined in a later 1973 version can hardly be overstated.[16] As Mayor Atcher himself noted at the time, "No large city in the world has ever had a proposal of this magnitude, to say nothing of the suburbs."[17] The plan saw the Outer Planets development as the center of a catchment area stretching 500 miles in radius and encompassing no less than 63 million population.[18] A clever salesman, Romano had pitched his Outer Planets development, or "Northwest Loop" (a reference to Chicago's downtown business and commercial area, the Loop) of five related developments—the Outer Planets residential blocks, an office plaza, an international center, the "shopping bag," and a "sports world"—to chime with Mayor Atcher's aspirations for Schaumburg, arguing that "Schaumburg will be the hub of the future metropolis."[19]

Unlike the sorts of low-rise detached residential housing, shopping malls, and office parks that were springing up around similarly accessible locations on the fringes of many metropolitan areas across the United States, the plan was for a distinctly high-rise form of suburbia. Some sense of this is afforded in images reproduced from the plan (see figure 7.3). Thus, as a local attorney recalled, the scheme probably did not fit well with the existing vision for Schaumburg. However, it was granted planning permission since "on the other hand there wasn't much here and it was a very attractive proposal. . . . Lee was a salesman. We were anxious for development. Woodfield had not opened yet."[20]

The alternative suburban future that the Outer Planets represented was clear from the introduction to the plan itself: "Its primary objective is to provide an exciting attractive alternative to the dull and formless environment now prevalent in contemporary American residential communities." Instead, then, the Outer Planets concept presented a picture of urban density in what is one of the lowest-density outer suburbs in the region. According to accounts of the day, it was considered financially viable, while its tall buildings were said, at least by the developer, unlikely to present any

Figure 7.3
The Outer Planets Vision of Suburban High-Rise Residential Buildings. *Source:* Village of Schaumburg Library.

problems in terms of aviation at the nearby O'Hare Airport. However, history was not kind to Romano's plan. None of its elements were built, and by 1975 financial difficulties had seen zoning approval lapse, and the plan faded from view entirely. In some ways, however, the plan itself changed the history of this particular suburb. The scale and form of the proposal were certainly alien to the growing residential community. There was also growing unease over the long-run fiscal implications for the municipality, which would likely to have to assume responsibility for the maintenance of the infrastructure and utilities of PUDs. Both these concerns appear to have played some part in Mayor Atcher's eventually relinquishing his mayoral position in 1975.

Shorn of some of the architectural tastes of the late 1960s and early 1970s, the plan's vision of density, public transit, and the coupling of employment and residence is one that has begun to resurface as some of the social, economic, and environmental limits to the present suburban pattern have emerged. The impetus to address some of these limits is dimly visible even in a place like Schaumburg, though presently they remain intractable there politically and technically. This past vision of the future may yet, then, have something to tell us of a *post*-suburban age still to come.[21]

Schaumburg and New Regionalism in Chicagoland?

In the case of Schaumburg and the Chicago metropolitan region, our starting point must be a consideration of the relationship between local government and the prospects for regional-scale governance arrangements since, David Hamilton writes, "The Chicago region is noted for its political fragmentation and central city/suburban antagonism. . . . it has the most local governments of any metropolitan area in the nation. The city dominated the region politically as well as economically for most of its history."[22] Counties have been left as residual entities unable to suture together a proliferation of separate communities even through the likes of service provision. They have, however, encouraged the formation of intergovernmental or council of governments (COG) arrangements. Developments in the institutional makeup and policy advocacy in the Chicago metropolitan area (such as *Chicago Metropolis 2020* and its sequel, *GOTO 2040*) have also been regarded as informed by "new regionalist" thinking.[23] The environmental consequences of suburban development in the region have been significant and have added to the scope and scale of issues surrounding the financing of and provision for collective consumption. Despite evidence of a new and important measure of multitier governmental and nonprofit organizational cooperation on several regionally important issues, such as housing, economic development (including the O'Hare Airport expansion), transportation and planning agencies, and surface transportation, these characteristics continue set the tone for the limited geographic reach of any post-suburban politics in which individual communities like Schaumburg are embedded.[24]

While Chicago's post–World War II suburbanization quickly yielded initiatives to increase governmental coordination and impose some order on metropolitan expansion, bodies such as the Northeastern Illinois Planning Commission (NIPC) and the Chicago Area Transportation Authority had to contend with a political divide between the Democratic Party–dominated city of Chicago and the many Republican-leaning suburban communities, a divide that amounted to a cultural war.[25] Mayor Richard J. Daley (serving from 1955 to 1976) formed a powerful civic alliance dedicated to protecting the downtown Chicago Loop in the face of suburbanization.[26] O'Hare International Airport, located in the suburbs, has continued to occupy a central place in Chicago mayoral politics and the city-suburb relationship, insofar as Chicago annexed the airport and a narrow access corridor to it so as to capture its fiscal contribution for the city. Partly as a result, the NIPC's 1967 *Diversity in Order* regional plan, which advocated concentrating suburban development in fingers along existing rail lines, had little influence on actual development patterns.[27]

Some of the difficulties of instituting an effective new regionalism in the Chicago metropolitan area reflect the fact that its "suburban region is composed of multiple independent political entities, with different and often conflicting agendas and degrees of control over the physical environment and different potential urban fortunes."[28] Moreover, despite the suburbs' independence, the limited liability and horizons of political leadership reflect a structure of government in which mayors rely significantly on their village managers for vision.[29]

This city-suburb divide continues to exist. As one interviewee said, "Even given the different size and makeup of the communities, there is definitely a suburban thread. I don't know if it's for years the city was the big bad wolf and suburbia had to hang together and make sure that everything didn't get sucked back in?"[30] The social, political, and cultural divide between city and suburbs was underscored by the greater measure of self-organization and collective voice at the Illinois state level, the result of nine separate COGs formed by various geographic groupings of suburbs across the Chicagoland region.[31] Larry Bennett: "The evident success of more localized, sub-regional cooperation via the area's nine councils of government indicates that a limited regionalist perspective has already taken root in Chicago's suburbs."[32] Moreover, intergovernmental self-organization and the voice of the suburbs have had other effects. "The largest geographically defined delegation in Illinois State legislature is now the suburban Chicago contingent," writes Bennett. "In the future, Chicago mayors will find it increasingly difficult to solve their problems by 'going over the head of the suburbs' to deal directly with Illinois governors. Cooperation—from the ground up—is likely to be forced on Chicago mayors as the necessary principle in governing city-suburban relations."[33] Reflecting on this emerging influence at state level, one interviewee noted, "Most of the suburbs out here are politically very competitive, which can be an advantage, because we have Republicans and Democrats in the legislature. And that is sometimes a lot better than having one party, because if your party is not in control . . . you tend to get less. . . . So you usually manage to get help from the leadership in getting stuff done for us."[34]

Little wonder, then, that former Chicago mayor Richard M. Daley moved to dampen some of the "city-first" focus of his father, Mayor Richard J. Daley. This included the creation of the Metropolitan Mayors Caucus in 1997 to bring together the city and mayors balkanized into nine separate COGs covering the greater Chicago area, with the original economic focus now having broadened.[35] Notably, this effort began with outreach to the Northwest Mayors Conference (NMC), of which Schaumburg is a part, owing to the economic strength of this grouping of municipalities.[36] Affirming some scholarly accounts, interviewees noted that the contrast between father and son as mayors could hardly have been more different as regards their attitudes to the suburbs.[37] However, others have argued that "beyond Chicago's city limits, Mayor Daley's record has verged on the schizophrenic," so much so that he failed to consistently pursue "a politics of regional and inter-governmental collaboration," in Bennett's words.[38]

Mayor Richard M. Daley's tenure also witnessed other examples of institutionalization at the regional scale. Notably, the Commercial Club of Chicago's *Metropolis 2020* plan sought to create a discourse that bound the city and suburbs when identifying a series of impediments to future metropolitan area growth. Although controversial in reigniting suburban concerns of city-centered ambitions, *Metropolis 2020* is widely considered to have prompted the rationalization of combining planning and transportation bodies into a single metropolitan-scale transportation and land-use planning body, the Chicago Metropolitan Agency for Planning (CMAP). In turn, CMAP has continued to promote a discourse and policy agendas centered on addressing metropolitan-wide inequalities in health and education, the lack of affordable housing, poverty and racial segregation, and even equalization of tax revenues, though so far these ideas and agendas have gained little traction in a place like Schaumburg. CMAP's work here has also been aided and abetted by NGOs such as the Center for Neighborhood Technology and the Metropolitan Planning Council (MPC), which have focused on regionwide housing affordability, with the former in particular emphasizing the role of transportation costs in overall affordability.

Local Growth for Collective Consumption in the Chicago Region's Post-Suburb

The coordination challenge facing the existing regional institutions such as the nine COGs, the Metropolitan Mayors Caucus, and CMAP in generating recognition of the hurdles to financing the sorts of collective consumption needs and expenditures that exist across the metropolitan area is big. Moreover, intergovernmental cooperation among the suburban communities needed to address those issues is also hampered by the limited liability felt by the citizens and politicians of Schaumburg.

The Post-Suburb and the Suburban Region

The interrelationships between Schaumburg and its neighboring communities in the northwestern suburbs of Chicagoland have been apparent from early on. There have been proposals to combine Hoffman Estates and Schaumburg. Virginia Hayter in her first year as mayor of Hoffman Estates reached out to Schaumburg mayor Robert Atcher to discuss merging of the two communities. After all, as the *Hoffman Herald* newspaper opined, "Every property owner in Schaumburg has profited financially by the creation of Hoffman Estates. . . . Any industry which comes to Schaumburg Township in the near future will come primarily due to the already established Hoffman Estates and the labor supply which is here or will move here."[39] However, Schaumburg's Mayor Atcher remained opposed to a merger, with Hayter later describing him saying at the time, "Fine, but the price of admission is $10 million." Atcher wanted to cover the cost of extending services to Hoffman Estates and demanded that Hoffman Estates upgrade its roads and water system before he would agree.[40]

An account by Cheryl Lemus of the development of Hoffman Estates doubtless represented the reality for Schaumburg and the northwestern suburbs as a whole by the 1980s when suggesting that "clinging to the notion that the suburb was a bedroom community was no longer acceptable. Hoffman Estates had become part of the Golden Corridor, a group of suburbs from O'Hare Airport to Elgin that integrated their location on the northwest tollway."[41] In recent years, although the interdependencies among the northwestern suburban communities have been recognized in a COG, some of these communities have continued to cast envious glances at Schaumburg's economic base and its resultant low or nonexistent residential property tax. Here the fact that Schaumburg's self-sufficiency (or what, in chapter 3 and 4, I dubbed its post-suburban politics mark I) rubs up against regional and subregional politics (or what I have called post-suburban politics mark II). While the view held in some neighboring communities, namely, that interdependence among the northwestern suburban communities means that Schaumburg's growth could not have occurred without housing nearby, this is not the view held by Schaumburg leaders. Instead, they feel that the other communities build off Schaumburg, which is the economic engine.[42] A specially convened Regional Tax Policy Task Force to inform CMAP's board has called for proposals for reform of the system of state redistribution of taxes, arguing that it "does not provide support for major regional needs that cross jurisdictional boundaries, such as transportation infrastructure."[43] Suggestions for reform cover how the sales tax is redistributed, since 26 percent of the metropolitan region's population receive 50 percent of all municipal sales tax disbursements.[44] The mayor of Schaumburg has been particularly keen to defend the status quo in these respects as, according to one interviewee, while overall there would be more winners than losers across all communities, Schaumburg would be the biggest loser.[45]

In this respect, Schaumburg's initial vision of itself as the capital city for the northwestern suburbs means that it is a region unto itself. As one interviewee said, "It definitely does have more of a regional feel to it. It is definitely a center within the northwestern suburbs. . . . It is viewed as a supramunicipality, more than just a town; it is a region that spills over into Hoffman Estates and Rolling Meadows." As a result, Schaumburg has tended to remain aloof from a number of intermunicipal cooperative projects. The MPC has been working with five communities in the northwestern suburbs neighboring Schaumburg on housing-related issues, but overtures to Schaumburg on this project and others have been unsuccessful.[46] As the same interviewee surmised, Schaumburg, through its mayor, Al Larson, has its own way of doing things and is sufficiently powerful and connected to avert becoming involved.

TOD Denied: STAR Line Fails to Materialize

Schaumburg sits in a "Golden Corridor" of suburbs to the northwest of Chicago, many of which owe their development to the access afforded by I-90 and are integrated with Chicago to a lesser extent by historical rail lines. The reworking of this suburban expanse is a huge challenge to be faced if many of the contradictions of a suburban fix are to be addressed. As one interviewee said, "Retrofitting the whole office corridor along I-90 to try to make it more supportive of transit and also to bring more affordable housing in proximity to jobs, this would be a high priority because it could have regional impacts. The northwestern corridor of Cook County has seen an employment boom over the last ten, twenty, maybe thirty years that has really shifted the jobs focus in a major way away from the central city, and all of that employment and economic development has occurred without affordable housing development in proximity. So it has aggravated the jobs-housing mismatch in the region."[47] The arrival of public transit infrastructure is central to the possibilities of reworking suburban space and the managed growth objectives that have characterized the NMC. Not surprisingly, then, the NMC, which includes Schaumburg, has been instrumental in bringing forward initiatives for suburban mass transit, though as we have already seen, the financing of such fixed infrastructure is critically dependent on proving ridership and hence value for money in advance.[48]

The so-called STAR (suburban transit access route) line extension of the Chicago metro system out through the northwestern suburbs using the central reservation of the I-90 is one major proposed infrastructure development that could drive a greater density and mix of development in these suburbs, including Schaumburg (see figure 7.4). It would have been revolutionary in being the first suburb-to-suburb mass transit rail line anywhere in America and was an initiative in which suburban communities were involved from the outset. Unfortunately, unlike the extension of the Washington, D.C., metro out through Tysons Corner (discussed in the preceding chapter), for the foreseeable future it remains on the drawing board. While the planned extension has suffered from the same federal strictures regarding funding that all mass transit schemes face, it is perhaps ironic that the lack of existing density of employment uses in Schaumburg's regional commercial center may have exacerbated this particular issue. Unlike the metro extension through Tysons Corner, it also did not have the nearby Ballston-Rosslyn corridor to draw on as a model. Instead, the uniqueness of the STAR line proposal and the lack of any comparable suburb-to-suburb line have been factors causing delays in funding.[49] Unlike in the Tysons Corner case, business does not appear

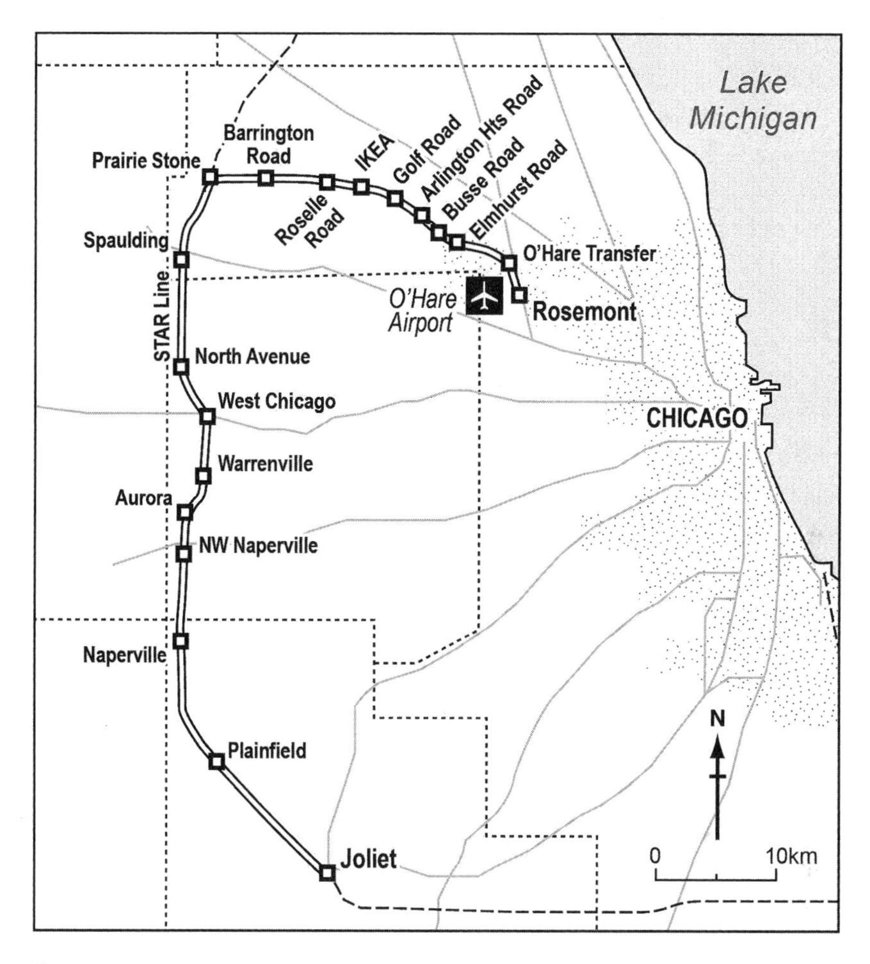

Figure 7.4
The Planned Route of the STAR Extension of the Metro System

to have been especially involved in lobbying for the STAR line. Business involvement in regional planning and politics continues to remain more focused on Chicago itself, while the organized lobby that does exist for high-speed rail likely has divided attention regarding the development of suburban rail lines such as the STAR line.[50] Where business interests do exist, they appear to have been relayed through the mayor of Schaumburg's political support for the STAR line. Finally, as one interviewee noted, Illinois was until recently one of a handful of states that did not have existing legislation that enabled private-sector contributions to a scheme like the STAR line and that proved so important to the development of mass transit through Tysons Corner.[51]

The obstacles to developing the STAR line have been not merely financial and technical but also reflect what some perceive as a lack of appetite for a greater density of development in the suburbs as a whole, the result of enduring preferences for convenience and even the practical difficulties with mass transit in Chicago's winter, but also a consequence of the political differences among the communities among the northwestern suburbs, for the STAR line would integrate suburbs of very different character.[52] One interviewee noted how "In many suburban areas there is a resistance to density. There is really a conflict between needing more density in order to provide transit and to use less land and all the benefits that can come from that and what is perceived as increasing traffic—and it is not just perceived it does."[53] To a certain extent, some of these disagreements have been moderated by the fact that the northwestern suburbs form a relatively cohesive set of interdependent settlements. In this respect, "the mayors along the line saw the potential for both the overcrowding on the I-90 corridor constrained growth in the western suburbs and the disconnect with the transit system to tie the housing-rich areas with the job-rich areas—Schaumburg being the prime one of those."[54]

However, some suburban communities, collectively referred to as "the Barringtons"—a group of primarily very low-density residential communities at the western edge of the STAR line's reach—have their own separate COG and have been opposed to proposals. The oldest community of Barrington, while already developed around a railroad and conceivably standing to benefit from a greater density and mix of development of its downtown, has not wanted to be bisected by more train traffic. South Barrington is characterized by extremely low-density "McMansion" houses (see figure 7.5, which shows those in neighboring Inverness). According to the former mayor of Hoffman Estates, the residents of the Barringtons have simply not wanted development, having formed their own COG at least in part to fight any proposed developments on less than 5-acre lots, and have fought the widening of Barrington Road and any sort of transit system through the northwestern suburbs. So Schaumburg and Hoffman Estates rub up against this position.[55] The story is complicated by the fact that the Elgin, Joliet and Eastern (EJ&E) freight line, which was to form part of the STAR line route, was acquired by

Figure 7.5
The Splintering Urbanism of Outer Suburbia: McMansion houses in Inverness. *Source:* Author's photograph.

Canadian National Railway to run freight. In light of the problems presented by Canadian National in guaranteeing access to the line to run STAR line trains, foreign ownership of the line therefore generated opposition to the acquisition. However, it also cleaved an opening, into which opposition to the STAR line entered, in the guise of concern over the foreign ownership of public assets.[56]

The STAR line project has been supported strongly by current Schaumburg mayor Al Larson through the Northwest Municipal Conference. There is certainly a sense that Schaumburg's interest in the STAR line is motivated by a desire to further strengthen its fiscal base through a better and more intensive use of land at the northwestern boundary of its regional commercial center. This is itself partly a function of the legacies of low-density development in this part of the business center. This part of the regional commercial center is also home to a new convention center funded by the Village of Schaumburg and aimed at retaining local expenditures in hotels, restaurants, and shops elsewhere in the regional commercial center. With a view to leveraging the potential of the STAR line to underpin a greater density of and mixing of uses within developments in the northeastern part of its regional commercial center, the Village of Schaumburg considered it suitable for TOD and tax increment financing (TIF) to capture the uplift in land values to finance infrastructure improvements, mainly in the form of access ramps to I-90.

Optimism over the potential of the STAR line to drive TOD briefly sparked cooperation with the neighboring blue-collar community of Rolling Meadows to explore a more rational set of land parcels and road layouts. Some indication of the appetite for these changes is that the MPC, with its emphasis on housing affordability issues, was assisted by the Community Building Initiative of the Rolling Meadows Task Force. The key concerns were to avoid displacing existing residents and to preserve levels of affordable housing as the municipality and owners and developers sought the opportunity to redevelop a low-density, 36-acre area of 692 two-story garden apartments, since this site represented fully 30 percent of Rolling Meadows' affordable housing stock.[57] The preference at the time in the proposals put forward by consultants advising the task force, The Lakota Group, would have seen the area developed into 1,584 residential units, including several twenty-story buildings, and 415,000 square feet of commercial space in a mixed format.[58] However, in the economic downturn of 2008 and the consequent "significant drop in the Equalized Assessed Valuation of the TIF District, the Village terminated the TIF District in 2011."[59] Studies on the potential of this area continue, though the case for a TIF district appears now to be couched in terms of "conservation," or addressing blight in the area.

While there have been a number of New Urbanist–inspired mixed-use developments of some density elsewhere in the Chicagoland region, the MPC has been wary of being locked solely into a New Urbanist framing of new housing development. The same reluctance to frame possibilities for a greater density of development in New Urbanist terms is also evident among Schaumburg planners, who prefer to talk of transit possibilities, possibly because of a lack of comprehension of the term "New Urbanism" among, and the sensibilities of, Schaumburg's village board, given how different such developments typically look to the rest of the existing community.[60] Nevertheless, the language used in a recent draft concept plan revisit of this part of the commercial center of Schaumburg is imbued with much of the regional planning sentiments of CMAP and its predecessor organization, and continues to express a desire for TOD. This is a part of Schaumburg where "the functional and economic obsolescence of existing uses no longer meet the needs of 21st Century companies . . . nor do they respond to the market opportunity presented at one of the most desirable and accessible locations within the Chicago market."[61] The draft plan for a northern section of the Woodfield Regional Center depicts "Walden Village" in term of "the desire of the Village [of Schaumburg] to establish the TOD area as a fully integrated mix of mutually supportive land uses, open spaces, trails, roads and transit services. . . . [to] provide for a more sustainable community where people can live, work, shop and find high quality entertainment and recreational opportunities within a walkable environment."[62]

Legacies of Corporate Capitalism

Shopping malls and corporate campuses have invariably made their own distinctive contribution to America's suburban morphology with their dedicated parking structures. They also remain perhaps the hardest of suburban spaces to rework because of the entrenched private-sector interests involved and the scale of the investments in suburban formats of development.

Some have argued that the suburban mall economic model has reached its limits, such that malls now present major opportunities for redevelopment in many suburbs.[63] These debates touch on the Woodfield regional shopping mall, which plays such an important role in Schaumburg's fiscal health. The future of these regional shopping malls is open to question, not least because it may be coupled with questions of greater density of development and population. As one interviewee said, "I have a 'tiger theory'—that regional shopping centers are like tigers. They are an endangered species and there are just so many of them. And, once they are located, people wake up to the fact that tigers eat meat. And in the case of regional shopping centers they eat density, and so you have to increase density to keep them viable."[64]

However, mall operators themselves are tenacious defenders of the suburban, automobile-oriented business model, including in Schaumburg. Some indication of this is provided by the fact that Woodfield Mall sued the developer of the One Schaumburg Place (now Streets of Woodfield)

Figure 7.6
Parking Lots Surrounding Offices and Woodfield Mall in the Regional Commercial Center Schaumburg. *Source:* Author's photograph.

development in 1984 over increased traffic generation, with the Village of Schaumburg countersuing Woodfield Mall under antitrust law.[65] Parts of One Schaumburg Place were torn down after six years to make way for the opening of Streets of Woodfield. The conflicts here reflect the views of mall-operating companies and managers, which as yet do not detect a weakening in the market for the regional mall. As the manager of Woodfield Mall observed, "I think [for] the suburban markets, especially in the Midwest, the term "pedestrian feel" is not really accurate for retail developments. Although there are attempts to take it in that direction, the customer is still demanding accessible parking and convenience. Over the years the customer demand has evolved toward choice . . . as opposed to proximity."[66] Even within the regional commercial center itself, the spatial separation of different developments is considerable, to the point that public transport is not cost-effective to run *within* the center, let alone to and from the regional commercial center and residential areas (see figure 7.6). As an interviewee from a local commercial bus operator said, "You have a basic problem of there's a non-grid. There is a tremendous amount of sprawl, a tremendous area to be served. There is a disconnect between retail and other commercial areas. The tremendous size of it to begin with doesn't fit well with any kind of fixed route service."[67]

A rash of corporate campuses appeared in the suburbs from the 1960s on. The nationally and internationally dominant—often monopolistic—position of these corporations saw them erect expensively designed and appointed buildings on extensively landscaped campuses. However, as these American companies have faced increasing competition in a post-Fordist era, the return on these assets looks increasingly questionable. Companies have sought to downsize, and to take advantage of the competitive bidding that exists among suburban residential communities searching for a larger tax base. The major investments sunk in these corporate enclaves mean that they represent significant impediments to the reshaping of suburbia.[68] Yet, notwithstanding their lack of connectivity to mass transit, when abandoned, they may also represent major opportunities for integrated mixed-use developments that would go some way toward reshaping the suburban form and some of its environmental impacts, not least because they entail the sorts of large single-ownership land parcels that are rarely found in metropolitan areas. In Schaumburg's case, Pure Oil Corporation made way for the building of a university campus and an IKEA retail development. Motorola remains on its campus, but a recent report indicated that 35 percent of the office space on the campus was vacant (see figure 7.7).[69]

Growth versus Environment?

The environmental costs of suburbanization across the Chicago metropolitan region have been recognized by regional planning institutions such as CMAP. CMAP's *GOTO 2040* observes that "development over the last several decades has resulted in a pattern of land use that is not sustainable. Development in the last half of the 20th Century has overall been a story of outward expansion, consuming vast amounts of land and requiring huge investments in water, wastewater, and transportation infrastructure."[70] What is also clear is that CMAPs forecasts suggests the Chicagoland region will grow by around two million people by 2020 but that infill is likely to accommodate only a fraction of this population growth.[71] So that

Figure 7.7
Motorola's Corporate Campus in Schaumburg. *Source:* Author's photograph.

whatever sequel to suburbia is in the making, it will be a combination of traditional suburban development and infill that reflects the diversity of suburban communities.

The lack of concern for the environmental impacts of suburban development in Schaumburg and elsewhere across the outer suburban communities in the Chicago metropolitan region continue to reflect both the limited liability associated with governmental fragmentation and the historical animosity between the city and the suburbs in the metropolitan region. For the northwestern suburbs of Chicago, the collision between growth and environment is one that remains firmly imbued with suburban ideology associated with the search for high residential amenity in the form of low-density development and governmental secession and home rule. Grassroots politics appears to be particularly thin, and there is very limited resident participation in big-issue planning ideas. There is little or no glimpse here of environmental issues driving regional-scale intergovernmental solutions. This is the context in which the present and near future possibilities for a reworking of suburbs through the likes of New Urbanist–style or TOD developments must be understood.

The Chicago metropolitan area is more varied than Miami in terms of its existing built form and density and associated modes of transportation. The metropolitan area includes the northwestern suburbs, of which Schaumburg is a part and where a mix of different eras of urbanization and modes of transport accessibility can be found. Here, then, there are some signs of transformation. As the program director of transportation and community development observed, "As you look at the Chicago region, there are a cluster of communities in the northwest part that have all implemented some degree of transit-oriented development. Arlington Heights is one of those, Des Plaines is another. . . . My perception is that Schaumburg is a different suburb from these others. The regional shopping centers overwhelm, or they are the signature image that people have of Schaumburg, and that is not true of those other communities. And the other communities are smaller. So you can have a bigger impact with a smaller retrofit. To make a difference in Schaumburg, it would have to be a redevelopment of a pretty dramatic scale."[72] The most notable instances of this reflect tendencies toward the greater density and mix of development at those historical communities centered on the railroad. Arlington Heights has become one celebrated example, though even there residents mounted significant opposition to developments in the center of town, such that the mayor noted the political courage required even when redevelopment was not encroaching significantly on residential areas.[73] In Arlington Heights and elsewhere, a measure of decline or blight has quickened interest in TOD.

The Limited Liability of a Mark I Post-Suburb

Any political tensions over the balance to be struck between supporting economic and population growth and protecting the environment are less apparent in the case of Schaumburg. As Schaumburg grew, "the juxtaposition of farmland and the emerging Schaumburg skyline was typical of the view one might see in driving through the village during the 1970s and 1980s," though this rarely appears to have drawn debate among the population as to the environmental costs of such development.[74] Instead, environmental politics continues to center on local residential and environmental amenity, as the same account of the growth of Schaumburg implies. "The collision between town and nature came to a head in 1971 and has taken shape over the past decade. Open space has become almost non-existent in today's suburbia."[75] As the mayor of Schaumburg's adjacent community, Hoffman Estates, observed, "I'm not sure there is a big movement to make suburbs more dense. We have become denser over the years because we have gotten a lot bigger. People like having a choice. Actually living in an urban environment isn't a choice here right now because we lack the transportation."[76] With a current population of approximately 76,000, Schaumburg is largely built out. As Schaumburg's original and still serving attorney observed, "Schaumburg is no longer the new kid on the block. So it's altogether possible that there will come a time in the next ten or fifteen years when redevelopment will become necessary. The only kind of redevelopment existing we are getting now are teardowns. But as compared to Evanston and Arlington Heights, Schaumburg still doesn't need that kind of redevelopment."[77] As another interviewee confirmed, even though Schaumburg is now essentially built out and there are opportunities for redevelopment, very few have actually happened so far.[78]

Yet Schaumburg politics is also doubtless not as obstructive to such possibilities as that evident in those very low-density and wealthy communities farther out from the city, such as the Barringtons. Here exclusionary zoning ensures that much of the population is housed in developments off the fixed infrastructure water and sewerage networks. Here housing and lifestyle preferences and politics continue to reflect the bourgeois ideology of suburban retreat from the city and the preservation of local residential and environmental amenity. Indeed, the Barrington Area Council of Governments formed as a BACOG primarily to oppose a major development of 18,000 homes in South Barrington.[79] The residents' concerns have since grown to encompass issues of water recharge. In this setting, increases in population and greater densities of development have significant implications for the environment, and in particular for the recharge of the shallow aquifer from which many exurban and rural communities draw. As the mayor of Barrington explained, "Water is a big issue for the region. Schaumburg gets Lake Michigan water. All Barringtons get their water from a shallow aquifer, and the low density allows recharge of that."[80] Here an environment-versus-growth dilemma is apparent and has resolved itself locally in favor of the environment, though it is also inextricably linked to a secession of exclusionary residential communities that is contradictory to the sorts of wider regional institutional dialogue and arrangements needed to protect and secure water supplies. The communities of the BACOG are joined as the eastern extreme of wider swath of exurban and rural counties and incorporated communities across five COGs and five counties into the Northwest Water Planning Alliance.[81]

However, the challenge of fashioning a degree of density is a function not just of the lack of existing mass transit infrastructure and options but also of the contradictions of the wider systemic planning for the automobile, which include notably extensive land-use separations. These separations have been extreme in a place like Schaumburg, where such forethought was given to them. To underline this sense of land-use separation and the low density of development across Schaumburg, figure 7.8 provides a schematic illustration of the relative sizes of the three development areas compared in this book. As one expert involved with sustainable transit-oriented urban planning summarized then, "of the challenges of retrofitting Schaumburg . . . the two biggest are the expanse of land that it covers. It is sprawl upon sprawl. I would say a second major hurdle is the fact that the regional shopping centers are themselves so massive and numerous.

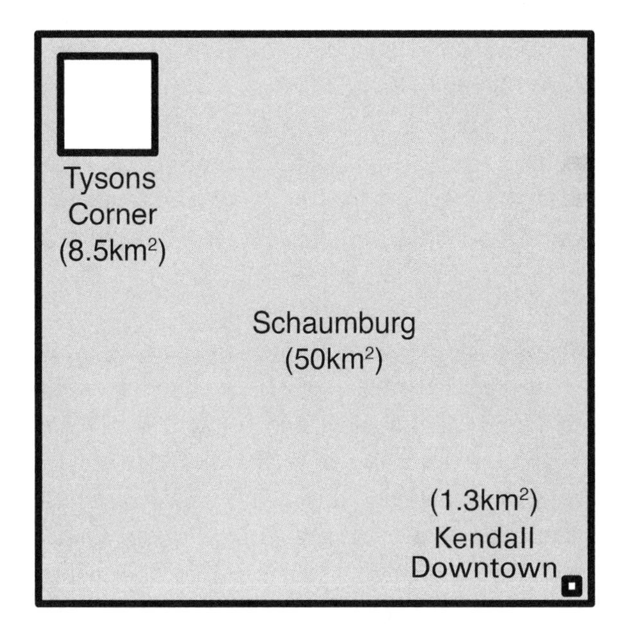

Figure 7.8
The Relative Scales of Kendall Downtown, Tysons Corner, and Schaumburg

I wonder what the ability is to retrofit enough to make a difference and have an impact."[82] While there is a measure of flexibility in public transit services, and there is a tendency to think that the likes of bus services are flexible enough to adapt to serve this suburban morphology and separation of land uses, suburban governments have rarely proved themselves at all elastic in their land-use planning responses in ways that might be more suitable for the supply of public transit. As the same interviewee went on to say, "Working with Schaumburg and the other areas . . . they think that we can do so much for them with the public transportation system, but they make the public transit system so undesirable and unattractive and slow and inefficient because of the land-uses. . . . They automatically assume the transportation is going to come to them, and there is a giant disconnect of bringing the land uses coming to the transportation."[83] As a result, a more transit-oriented pattern of development in the suburbs is likely to be rather different from that aspired to. One interviewee referred to it as hybrid. "It is going to be a hybrid. From the suburban point of view, you are not going to force urbanism down their throats because they don't want it and they are rich enough to avoid it."[84]

It might be objected that the 50 km^2 covered by all of Schaumburg (residential and commercial land uses combined) is not strictly the correct metric to use in comparisons with Tysons Corner and Kendall Downtown. However, even the more modestly sized comparator of Schaumburg's Woodfield Regional Center (which does not include all of the industrial or commercial land in the Village) is 50 percent larger in area than Tysons Corner. Indeed, the Woodfield Regional Center is large enough to contain six Cincinnati downtowns, for example.[85]

Suburban Business as Usual? O'Hare Expansion and Western Access

Former Chicago mayor Richard M. Daley is generally acknowledged to have had to take the suburbs with him with respect to his ambitions to expand the capacity of and increase access to O'Hare Airport as a means of further developing the Chicago regional economy. O'Hare and its east access road fall within the jurisdiction of the city, and plans for its expansion involved taking land from adjacent suburban communities, so the increased externalities affecting them had to be sensitively negotiated.

The voice of a distinctly regional roads lobby was apparent in the planning deliberations for Tysons Corner, as we saw in the preceding chapter.

A regional roads lobby has perhaps been less overtly apparent in the case of the Chicago metropolitan area, but then again, road improvements have figured prominently in public debates to develop the potential of O'Hare Airport and have been promoted by the Illinois Department of Transport (IDOT) and the Illinois Highway Authority (IHA), and have had the organized support of city-based business. There is little doubt that access to one of the world's busiest airports is unusually constrained by one toll freeway entrance from the east. However, it is difficult to see how a new western access will not present the further proliferation of suburban densities of development further to the west of O'Hare (see figure 7.9).

The communities adjacent to the western side of O'Hare Airport are already built at suburban density, and it is difficult to see greater access promoting much in the way of greater density of commercial or residential

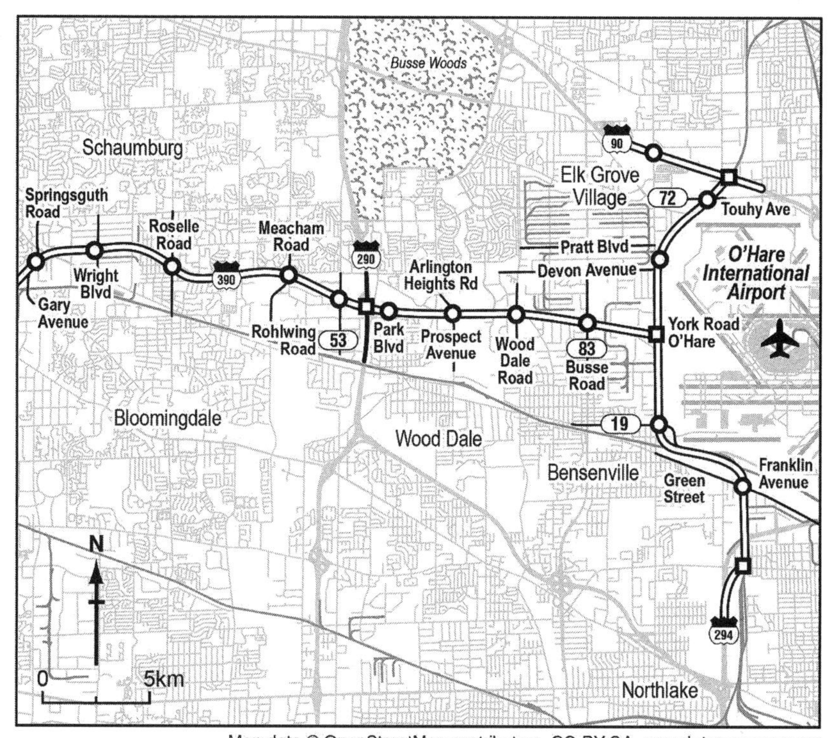

Figure 7.9
Western Access Improvements to O'Hare International Airport

development. Local resistance to airport expansion has focused on the broader environmental implications of airport expansion and on what, if any, opportunities access might provide for a reworking of the suburban format. Instead, one of the concerns has been the suburban land take in the name O'Hare Airport expansion. A Suburban O'Hare Commission formed of the suburban communities surrounding O'Hare Airport had opposed past proposals for expansion of the airport in light of the adverse environmental and amenity effects on their communities for two decades but have since gone on to support a western access to O'Hare.[86] The City of Chicago struck compensation deals with the two suburban communities most affected in terms of the land take and the environmental impact of Chicago's plans for expansion of the airport and increasing access.[87]

Conclusion

The northwestern suburbs of the Chicago metropolitan area form a region in which some measure of discourse and institutional development has emerged around the three key post-suburban political tensions noted earlier. The scale of the problem as it manifests in the northwestern suburbs in which Schaumburg is situated remains great indeed, especially in light of the expanses of territory and fragmentation of governmental jurisdictions involved and even the lack of public transit options in an otherwise historically transit-rich metropolitan region.

If a new urbanity has been coming along in America, as several commentators suggest, then, even more so than Tysons Corner and Kendall Downtown, Schaumburg represents this genuinely hybrid new urbanity—a city in function but not in form. It is far closer to the edgeless rather than edge city form of development that is now characteristic of the majority of contemporary developments in suburban America.[88] In the distinctive paternalistic politics driving a balance of employment and residential land uses, it was a community born post-suburban. It was from the outset to be a new kind of suburban city in which the need for local self-sufficiency was recognized. Yet in regard to its perfection of the suburban morphology and separation of land uses, it was planned with thoroughly modern sensibilities. One might say that the sheer suburban modernity of its form is something that is unlikely to be undone incrementally. In some respects, the legacy of modernism is stronger and more complete here than in the typical residential suburb, so it may require as many federal and state dollars to radically transform it as it did to unleash it in the first place. Brenda Scheer and Mintcho Petkov "do not expect that the edge cities now developing can be adapted to be more like traditional downtowns."[89]

Moreover, while self-government has been important to Schaumburg's recognition of the limits of traditional residential suburbia in its mark I post-suburban constitution, home rule may yet come to be an obstacle to a community finding itself in some of the regional-scale solutions to the contradictions of suburbia in a way that can manage the development of greater infrastructure for collective consumption, notably mass transit and the sorts of greater density it can drive.

In Schaumburg, the full elaboration of post-suburban politics is distant on the horizon and of little immediate local concern, yet it poses a challenge the community will have to address eventually, in concert with neighboring communities in the region and subregion. It is a distant glimpse of a future that suggests that much of suburban America will not be post-suburban anytime soon. The two signature issues in the northwestern suburbs recently, the failure of the STAR line and the expansion of O'Hare Airport, underline this slow plod toward a post-suburban future. The airport expansion in particular embodies a sense of "suburban business as usual," while the subsidence of the STAR line metro extension proposals seems to imply a denial of subregional-scale solutions to the mitigation of some of the environmental impacts of suburbs and responses to their collective consumption needs. In this respect, the contradictions of localist mark I post-suburban politics appear almost as significant as those of suburbia itself.

8 Conclusion: After Suburbia

I've thought about it a lot, and what seems to me to be really wrong with Shady Hill is that it doesn't have any future. So much energy is spent in perpetuating the place—in keeping out undesirables, and so forth—that the only idea of the future anyone has is just more and more commuting trains and parties.

—John Cheever, *The Country Husband*

Over the years, many terms have been used to try to capture the spatial expansion of the American metropolis and its different constituent elements. Some terms offered for this purpose include galactic urbanization, post-metropolis, edge city, technoburb, edgeless city, and boomburb. They have recently been joined by the expression sure to end all terminological proliferation—planetary urbanization. In this book I have settled for a term already in use for some time, post-suburbia. In part, this is because I have no desire to add to the terminological proliferation. And in part, I wanted to use a term that arguably focuses less on providing a literal or metaphorical description of the appearance or form of contemporary urbanization and more on the open-ended possibilities of future development. This desire to avoid a constraining lexicon in turn arose because some of the terms that have appeared in the literature to date do not travel well outside the national contexts in which they were first used.

If the future seemed unthinkable in the halcyon days of John Cheever's fictional residential suburb of Shady Hill, whose voice we hear in the epigraph to this chapter, in this book I have suggested, taking my cue from a range of observers, that some of America's suburbs have a dimly visible post-suburban future. However, unlike some of the authors whose works I have cited I have been interested in using the term post-suburb not to indicate a new era per se or to locate such settlements relative to suburbs (though these dimensions of time and space are hardly absent from the preceding discussion) but instead to signal a continuity between suburbs and whatever they may be transformed into. That is, I have been interested in how the same actors that shaped suburbia will, as a result of suburbia's contradictions, necessarily be prompted to experiment their way toward the making of at least a partial post-suburban spatial fix. The presentation and discussion of a class of post-suburban settlements alongside suburbs and cities is important since, as Ann Forsyth points out, "how urbanists, the press, and the public talk and think about suburbs shapes how they can see such areas being developed and redeveloped in the future."[1]

Researchers are often more comfortable directing their analytical attention toward aspects of change, toward gentrification and urban regeneration, and tend to focus more on the city proper than on the suburbs and their accompanying residential rootedness. Thus, "although it is well known that the political economy of postwar development in the United States has been predominantly suburban in character . . . only a handful of researchers have investigated in any great depth the emergence and characteristics of redevelopment-oriented political regimes in the suburbs."[2] Though Amer Althubaity and Andy Jonas's observation from 1998 remains largely true today, interest in the suburbs and in the process of suburbanization has increased quite markedly.[3] Perhaps it is "hardly coincidental that suburbia's history is being revised at the same time that its physical fabric is getting retrofitted."[4] And yet that history is an expanding but divergent history of very different types of suburbs with different pasts and potentially very different futures. This book is one contribution to what I believe is a wide horizon for investigation of suburbs and their potential sequels that has rapidly opened up to view. This book, then, is one among several "sequels to suburbia" that could be written. Here I have concentrated on the possibilities for the fuller urbanization of suburbia from various suburban starting points. Devoting some attention to a post-suburban sequel to suburbia is important, as this class of settlement speaks to ongoing debates in sociology, urban studies, urban geography, planning, and architecture. The post-suburbs will be of intrinsic interest to urban historians and sociologists since they are likely to embody new political sensibilities to set beside those having to do with urban and suburban ways of life and associated ideologies and politics. Moreover, for sociologists, some of the practical problems being addressed in these settlements and by these communities speak to the local and concrete materialization of a politics of a second modernity and any systemic change in the direction of a post-car era.[5] Post-suburbs are some of the best vantage points from which to observe what postmodern urban theorists regard as processes that have turned the city inside out.[6] For economic geographers the fortunes of post-suburban settlements may pose anew interesting questions regarding the nature of any post-Fordist, spatial fix.[7]

None of this, of course, says anything about the many suburbs that will likely continue to endure a decline, with land and property not being revalorized anytime soon, nor does it say anything about those suburbs that, in consequence of the serious population and economic shrinkage of major American cities, are likely to be largely or entirely abandoned. Some of these other trajectories of change across the contemporary American metropolis were broached in chapter 3 and are every bit as worthy of investigation.

In Frederick Jackson Turner's *The Frontier in American History*, the frontier has a double meaning: as a mobile spatial expression of different settlement eras, and as a *moment* in which institutions and societal values are continually remade. As Turner noted, "American social development has been continuously beginning over again on the frontier."[8] Something of this sentiment has also been expressed recently by Dolores Hayden, who interprets the suburbs as the hinge between past and future.[9] In this light, the evolution of suburbs and edge cities is worth exploring for what it can tell us about potentially significant transformations in societal values in processes of urbanization. The evolving nature of urban politics provides important glimpses of such changes in societal values, and suburbs and edge cities may be the most appropriate places to examine insofar as the powerful political and ideological mix that has driven the development of suburbia in the United States shows signs of having been altered in important respects. If the cases of Kendall Downtown and Tysons Corner in particular are anything to judge by, it is a local politics that now in many instances corresponds more closely to the sorts of growth management dilemmas posed by emerging post-suburban realities in the United States than the untrammeled pursuit of land-extensive patterns of development implied in the concept of a (suburban) growth machine, on one hand, or the exclusionary antigrowth politics typical of affluent residential suburbs on the other. A greater appreciation of this post-suburban realm presents an intriguing research agenda within urban studies.[10]

Some indication of the growing latitude of the contradictions of suburbanization and the state's implication therein is found in the increasing breadth of a fourth era of urban growth management policies in the United States.[11] This fourth era Chapin sees as having an agenda broadened further from the likes of smart growth and transit-oriented development to include longer-term and geographically expansive concerns such as climate change and food security alongside those centering on the resilience and sustainability of communities in economic development terms. Not only do these additional risks of a "second modernity"[12] give some indication of the latitude of the suburbs' contradictions, they also indirectly threaten to dilute the focus of civil society and private and institutional actors on the reworking of suburbia into post-suburbia since the ostensibly new and wider-ranging claims of suburbs will increasingly be set against the enduring claims of cities and the new claims of exurban or rural areas. Here, then, suburbia's in-between geography[13] may yet be its downfall, since the planning challenge of reworking suburbia may fall between two rather different policy targets. There is the distinct possibility that, if done badly, any flattening of the American metropolis (in which the distinction between suburban and urban becomes increasingly blurred) could prove the worst, not the best, of urban and suburban worlds.[14] In isolation, growth management measures may, paradoxically, do much to replicate Los Angeles' problematic dense sprawl elsewhere among America's newest metropolitan areas.

For all the inertia seemingly associated with America's outer suburbs, it is also clear that at least some will be remade in some form or other in the future. That process is likely to be slower than in contexts where urban containment has conferred greater value on many suburbs, old and new, inner and outer, or else has driven a measure of revalorization and redevelopment of suburban land and property. What Neil Smith described as the spatial see-saw effect of capital switching back and forth—out of and back into the city, for example—is a process of many decades.[15] Some indication of the timescale to be reckoned with is provided in the case of London. Inner London's population declined for seventy to eighty years from the 1920s on before strong population growth for a number of reasons prompted a significant revalorization of land and property there in the form of gentrification.

And this happened in a context of strong urban containment! Moreover, it seems certain that containment per se has not been the primary factor, especially when set against such other factors as financial industry expansion, significant domestic and overseas migration, and overseas property investment in renewed population growth in London.

Judged on this sort of timescale, a survey of America's sequel to suburbia will be a multivolume affair. That is to say, the development of a post-suburbia will almost certainly be highly incremental, piecemeal, faltering, stop-and-go. Yet for those same reasons it will be all the more visible.[16] The outer suburbs will not be absorbed quickly into the extant fabric and administration of cities so as to become almost indistinguishable from the city proper, as the older inner suburbs were. The task of integrating the outer suburbs into evolving metropolitan administrative or governance structures and of connecting them to urban infrastructures, notably mass transit, will be a very deliberate one and one reflected on at length, owing to the funding situation.

Past Pastoral?

The message of this book is a cautious one and is drawn from an examination of three post-suburban political tensions as they have manifested at three sites. These instances provide just the barest glimpse of the extent to which the contradictions of suburban development are likely to be resolved in a distinctly post-suburban politics over the coming decades.

The first political tension sets the pursuit of growth against the protection of the natural and built environments. The environmental consequences of suburban development can hardly be overstated, but they are difficult to grasp in policy and political terms since they are global in nature—literally so, in the case of climate change. They are global in the more restricted sense of being beyond solution by recourse to a purely local political and policy perspective. Because of the centrality of the search for environmental amenity within the bourgeois utopia of suburbia, the environment remains a concern in post-suburban communities, some of which are grappling with the problems of the remaking of nature lost more commonly associated with historical city cores. However, for the most part, post-suburban politics remains characterized by a limited liability inherited from residential suburbia.

The second political tension considered in this book was that between the promotion of growth through the private accumulation of capital and providing for the collective consumption needs that arise from that growth. Here it is possible to identify a class of post-suburban settlements precisely on the basis of their addressing this particular political dilemma. A pragmatic political response to some of the limitations of residential suburban communities began to emerge quite early in the history of post–World War II suburbanization. Those responses represented what I have termed mark I post-suburban politics and did indeed embody what Jon Teaford has described as a subtle rather than radical adaptation of the suburban ideal. However, they were self-interested responses that centered purely on the desire or need for greater local self-sufficiency. If the instances described in this book are anything to judge by, it is clear that there is a long way to go before this dimension of post-suburban politics evolves further into the global sense of place implied in what I have termed a mark II post-suburban politics.

The third post-suburban political tension relates to the scale or scales of intergovernmental arrangements that might be needed to address issues surrounding the growing income, employment, and fiscal inequalities across metropolitan areas, to supply the sorts of collective consumption needs that attend future growth through the private accumulation of capital, and to address the mounting environmental effects of suburban development. Here the instances considered in this book gave little indication of post-suburban politics having departed significantly from the purely local orientation of residential suburban communities. As yet, then, it is hard to see how the emergence of a class of post-suburban settlements alongside a variety of suburbs within metropolitan areas might entail a distinct political and policy-driving force toward intergovernmental cooperation. Aside from these practical and policy challenges involved in any remaking of suburbia, attending to questions of governing suburban and post-suburban localities can be an element in the sort of theory building that makes connections between multiple scales of government and politics and has been called for recently.[17]

It may be an enduring conceit of academic commentators that suburbs should change—indeed, must change. However, as Robert Bruegmann reminds us, the appeal of the suburban ideal is strong, even if it is not shared by some part of the population.[18] The suburban pastoral appeal remains very strong across America and is on the increase among citizens across the global south. From the perspective of its avid consumers, the market for suburban living is unlikely to dwindle to naught anytime soon. Although there is no little debate over the likely prospects for the urbanization of suburbia, this in itself gives some clue as to just how opaque the decision of where and how to live can be. To be sure, consumers of the suburban way of life exist in a market that has been significantly distorted by all manner of positive incentives such that they will continue to live under the fog of claim and counterclaim for some time to come. The industries that have produced suburbia have traditionally done little to lift the fog obscuring how any sequel to suburbia would have to be shaped: "Homebuilders, land developers, and marketing advisers are all constituencies that must be won over if the campaign against suburban sprawl is to succeed."[19] At one level, and given the sorts of time frames commonly used in the discussion, it is hard to avoid Pierre Filion's gloomy conclusion: "Present suburban dynamics are deeply entrenched and conditions needed for the vigorous interventions required to alter these dynamics are lacking."[20]

On the other hand, two of the communities examined here grew from past visions of suburban development that were vertical rather than horizontal. The original scheme for the development of the Kendall Downtown area (discussed in chapter 5) foresaw a mixed-use and high-rise mini-city. More startling from a present-day perspective is the planning approval given to the planned "outer planets" complex in Schaumburg. That the complex did gain planning approval and that those involved at the time recall the process as a matter-of-fact occurrence raise the specter of the doubtless many other schemes proposed in similar communities across the United States. Such instances represent forgotten suburban futures rather different from what we now know and understand as the suburbs that came to be. Here, then, John Gold quotes H. B. Franklin to capture how "when we look at a past vision of the future, what we see is the past and, in reflection, ourselves."[21] As a consequence, the history of Schaumburg's outer planets, and perhaps many other lost suburban futures, is one worth uncovering and telling anew.

Moreover, one irony is that the mass suburbia that has produced the system can scarcely be made over by another mass market. The market for suburban living is already saturated. As it continues to grow incrementally, one saving grace is that the industries that made suburbia will likely have to become more attuned to particular housing market niches. At one extreme these housing niches will, of course, embody the "vulgaria" that represent something of a gross parody of the original bourgeois search for the suburban pastoral in extremely low-density outer suburban or exurban living.[22] However, it is also highly likely that other affordable housing niches will have to be catered to in most middle-income outer suburbs.

A second glimpse of post-suburban politics and economics is provided by the fact that the state—which has been heavily implicated in the development of America's suburbs—continues to regulate and incentivize suburban development. However, the unintended effects of the state's own interventions are by now quite apparent. Thus, one hope is that the visibility of suburbia's contradictions will drive rather different state interventions going forward. Such interventions might be focused on the greater development of mass transit in the outer suburbs, though there is little suggestion as yet that any proposed schemes are likely to benefit from the sort of focus and funding (on roads and mortgage relief) that drove the massive outer suburban expansion in the first place.

A third glimpse of post-suburban politics shows the role of suburban bureaucratic and political leadership. Despite the rather sanguine tone of this book, a certain level of awareness—very keen, in many cases—and innovation in governmental arrangements was apparent among those political and professional actors involved in each of the three cases examined. Many of those whom I interviewed live and work in suburban America and are among the best and brightest. In some instances they are well known statewide, nationwide, and even internationally. A significant interest in transit-oriented and mixed-use development has been reported among political, government, and policy leaders in America's largest and fastest-growing suburban cities, the boomburbs.[23] Elsewhere suburban political leadership has become visible to varying degrees beyond suburban jurisdictions, and with differing consequences for the suburbs themselves. At one extreme, Rob Ford's Toronto mayorship may be unlikely to lead to much change in suburban business as usual. London mayor Boris Johnson's Outer London Commission may balance the suburban voice and development opportunities with those of the center in a finance-dominated world city. At the other extreme, former mayor Pedro Castro's refashioning of Getafe as a city from a bedroom community to Madrid has been based precisely on a pragmatic politics of collective consumption.[24] Of course, there is enough in the history of metropolitan politics to suggest that suburbs have been a source of votes fought over in enlargements of metropolitan authorities without themselves being the source or substantive subject of leadership. Whether the suburban polity amounts to more than one of retreat and distinction from the city and, indeed, the metropolitan region is an open question, though there is enough in the glimpses of post-suburban America presented here to suggest that local politicians have an emerging metropolitan sensibility and influence.

Steven P. Erie and Scott A. MacKenzie have contrasted the political machines of New York and Chicago with the bureaucratic machine of Los Angeles.[25] The former have been strong in their promotion of city cores against suburban fragmentation, though their success in this regard appears little different from the sort of metropolitan regional fragmentation often more commonly thought to characterize the Los Angeles metropolitan region. Although Los Angeles might be seen as more thoroughly modern because of its bureaucratic rather than boss-style leadership, neither model is likely to be especially successful in halting the systematic production of suburban sprawl without a broad-based appeal not only to the general public (as residents) but to the grassroots movements that now represent the many special interests proliferating in a second modernity. However, at this moment it is less easy to judge the likely success of grassroots movements to effect progressive change. While some of these groups have figured strongly in the recent emergence of progressive agendas in Los Angeles, including the sorts of issues that I have taken as the signature of a post-suburban politics centered on the likes of investment in mass transit and regional governance and growth management initiatives, it less clear whether Los Angeles will be model or exception.[26] Certainly there is hardly enough in metropolitan Miami's Kendall Downtown area to suggest that an inclusive and progressive cosmopolis will arrive anytime soon.

The American Suburb after the American Century

It is hardly surprising that a book written about the United States by a British-based scholar should have something of an undercurrent to it, an implicit question of America's relevance to the rest of the world. Without wishing to propagate a conceit underlying the origin of much extant urban theory, there nevertheless remains a good measure of truth in the thought that trends in America are in advance of those elsewhere. On the one hand, then, the relevance of the American suburban experience to the rest of the world is quite specific insofar as it has been regarded as legally exceptional in its promotion of the single-family detached house in what was until recently only a domestic phenomenon—and is only being exported in earnest for the first time now.[27] Noting the centrality of suburbanization to the dominant position achieved by the United States in the global economy, George Gonzalez writes, "In challenging urban sprawl, questions are going to be raised about whether the global economy can exist as we know it with a post-urban sprawl U.S. society."[28] Indeed, this much has been emphasized in the literature drawn on during the course of this book and in recent debates surrounding the future of suburbs. While there is some evidence that a peak in automobile use may have been reached in the developed world, the sizable differences in automobile usage between developed and developing countries promise a problem of carbon emissions and associated climate change on a much grander scale than at present.[29]

On the other hand, the questions asked regarding the global economy and environment will be much broader than Gonzalez suggests. Writing from the Canadian vantage point, for example, Jill Grant argues that "the American suburb may reflect a particular history and face unique problems, yet it has become an international signifier of urban malaise for some and modernity for others. Trends that begin in suburbs in the USA eventually come to influence suburban form in affluent neighborhoods around the world."[30] The story of suburbia and glimpses of its sequel in distinctly post-suburban politics and in the search for a post-suburban form therefore have important lessons for those parts of the global south. As Ann Forsyth explains succinctly, "The sheer volume of suburban growth expected in coming decades is unprecedented. It will pose substantial challenges for planners not least because the places where considerable growth will occur lack the high incomes and well established government structures that can simplify problem solving. Opportunistic suburban developments that lack infrastructure and links to the services of more established parts of urban areas are likely to remain the dominant mode of growth. . . . Retrofitting such opportunistic suburbs will be a key area for urban planning intervention around the globe."[31]

Of interest in this respect is the extent to which the framework of national and metro-regional accessibility that was pioneered in the United

States in the form of an interstate highway building program will be emulated elsewhere, notably across countries of the global south. Since such programs are likely to be bound up with the desire to project a national sense of modernity and driven significantly by regional development objectives, it would be churlish to expect nation-states not to undertake them. The issue is more whether the right legal and administrative mechanisms are in place so as to extract betterment from the process, not least as a way to underwrite major infrastructure and other collective consumption expenditures. In China, since land is ultimately the property of the state or the commune, in theory, surpluses from the land development process ought to be available to be recycled into local infrastructure development and provision for collective consumption. There is some evidence of rapid retrofitting of the outer suburbs of Beijing, which were initially served only by car.[32] However, here and elsewhere where land ownership is more mixed, "the differential value attached to what is 'formal' and what is 'informal' creates the patchwork of . . . space that is in turn the frontier of . . . accumulation."[33] In some of these settings the growth machine dynamic is harder to accept as some sort of necessary evil since it can, in many cases, amount to the worst instances of "accumulation by dispossession," with the resultant displacement of populations and inadequate or nonexistent compensation for businesses and properties repurposed.[34]

In the late 1800s and early 1900s a remarkable and broadly progressive era of North Atlantic policy exchange stimulated many of the advantages of cities that we now take for granted.[35] The potential for endogenous points and patterns of urban policy innovation across the global south seems clear.[36] Given what I have said immediately above about the prospects for local growth machine politics to center significantly on opportunities for accumulation by dispossession, the relatively weak bargaining stance of many American cities, suburbs, and post-suburbs may mean that they have rather less to contribute than European cities on questions of leveraging developer contributions for collective consumption needs. However, it is also clear that American cities, suburbs, and post-suburbs as repositories of accumulated knowledge and experience regarding municipal affairs can have a part to play in satiating what is an enormous appetite for knowledge of all sorts regarding municipal affairs among cities of the global south today. This is an appeal, it seems to me, to a possibly lost or forgotten better nature in American urban affairs since the knowledge sought from the global south covers the likes of appropriate metropolitan arrangements for governing and for supplying crucial services such as transportation, water supply, waste water treatment, and refuse disposal, and appropriate regulatory and planning tools for controlling urban sprawl. It is a tantalizing opportunity and one on which some part of American diplomacy after the "American century" might center.

The world will look on with interest as America debates, and experiments with, what is to be done about its outer suburbs. The outer suburbs have been a major domestic business in and of themselves and perhaps the best representation of American culture. As some of the dominant multinational suburban industrial complex companies it gave rise to have begun to struggle, American business abroad may come ever more to revolve around the suburban ideal that it has now begun to export in earnest to the global south.

Notes

Chapter 1: Introduction

1. M. Clawson and P. Hall, *Planning and Urban Growth: An Anglo-American Comparison* (Baltimore, MD: Johns Hopkins University Press, 1973).

2. See R. Harris, "Meaningful Types in a World of Suburbs," in *Suburbanisation in Global Society,* ed. M. Clapson and R. Hutchison (Bingley, UK: Emerald, 2010); R. Harris and P. Larkham, "Suburban Foundation, Form and Function," in *Changing Suburbs: Foundation, Form and Function,* ed. R. Harris and P. Larkham (London: Spon, 1999), 1–31; and G. Hise, *Magnetic Los Angeles: Planning the Twentieth-Century Metropolis* (Baltimore, MD: Johns Hopkins University Press, 1999).

3. J. H. Kunstler, *The Geography of Nowhere: The Rise and Decline of America's Man-made Landscape* (New York: Simon & Schuster, 1993).

4. R. A. Walker, "A Theory of Suburbanization: Capitalism and the Construction of Urban Space in the US," in *Urbanization and Urban Planning in Capitalist Societies,* ed. M. Dear and A. Scott (New York: Methuen, 1981), 383–429.

5. D. Harvey, *The Urbanization of Capital,* 2 vols. (Baltimore, MD: Johns Hopkins University Press, 1985).

6. Regulation theory suggests that the capitalist economy has grown in a series of long term periods of economic expansion regulated by an articulation between producers, consumers and state and non-state institutions. The Fordist regime of accumulation (named after Henry Ford's pioneering of mass car production for mass markets), of which the United States was the leading international center, is commonly considered to have lasted from the 1930s until the early 1970s. See M. Aglietta, *A Theory of Capitalist Regulation: The U.S. Experience* (London: Verso, 1979), and subsequent discussion in, for example, A. Amin, *Post-Fordism: A Reader* (Oxford: Blackwell, 1992).

7. See R. Beauregard, *When America Became Suburban* (Minneapolis: University of Minnesota Press, 2006), 5–6.

8. C. Walker and P. Boxall, "Economic Development," in *Reality and Research,* ed. G. Galster (Washington DC, Urban Institute Press, 1996, 13–27, 19) quoted in Beauregard, *When America Became Suburban,* 83.

9. J. Persky and W. Wiewel, *When Corporations Leave Town: The Costs and Benefits of Metropolitan Job Sprawl* (Detroit: Wayne State University Press, 2000).

10. N. A. Phelps, "Suburbs for Nations? Some Interdisciplinary Connections on the Suburban Economy," *Cities* 27 (2008): 68–76.

11. Harvey, *The Urbanization of Capital.*

12. Beauregard, *When America Became Suburban,* 180.

13. Ibid., 65.

14. D. Hayden, *Building Suburbia: Green Fields and Urban Growth, 1820–2000* (New York: Vintage, 2004), 11.

15. M. P. O'Mara, *Cities of Knowledge: Cold War Science and the Search for the Next Silicon Valley* (Princeton, NJ: Princeton University Press, 2005).

16. Harvey, *The Urbanization of Capital.*

17. Transit-oriented development refers to a mix and density of development that can be sustained in close proximity (usually within a half-mile radius) to mass transit infrastructure (see, for example, www.reconnectingamerica.org, accessed December 16, 2014). The smart growth movement is concerned essentially with the promotion of compact urban forms (see www.smartgrowth.org, accessed December 16, 2014). The New Urbanism movement was founded in the 1980s; the Congress for the New Urbanism, according to its website, dedicates itself to "the restructuring of public policy and development practices to support the following principles: neighborhoods should be diverse in use and population; communities should be designed for the pedestrian and transit as well as the car; cities and towns should be shaped by physically defined and universally accessible public spaces and community institutions; urban places should be framed by architecture and landscape design that celebrate local history, climate, ecology, and building practice" (www.cnu.org/charter, accessed December 16, 2014).

18. L. S. Bourne, "On Schools of Thought, Comparative Research, and Inclusiveness: A Commentary," *Urban Geography* 29 (2008): 177–186.

19. Beauregard, *When America Became Suburban.*

20. U. Beck, W. Bonss, and C. Lau, "The Theory of Reflexive Modernization: Problematic, Hypotheses and Research Program," *Theory, Culture & Society* 20 (2003): 1–33.

21. J. K. Galbraith, *The New Industrial State* (Boston: Houghton Mifflin, 1967).

22. M. Gottdiener, *Planned Sprawl* (London: Sage, 1977).

23. See G. A. Gonzalez, *Urban Sprawl, Global Warming, and the Empire of Capital* (Albany: State University of New York Press, 2009), 111; and P. G. Newman and J. R. Kenworthy, "Gasoline Consumption and Cities," *Journal of the American Planning Association* 55 (1989): 24–37.

24. Persky and Wiewel, *When Corporations Leave Town.*

25. J. Teaford, *Post-Suburbia: Government and Politics in the Edge Cities* (Baltimore: Johns Hopkins University Press, 1997).

26. See L. H. Masotti, "Prologue: Suburbia Reconsidered—Myth and Counter-Myth," in *The Urbanization of the Suburbs,* ed. L. H. Masotti and J. K. Hadden (Beverly Hills, CA: Sage, 1973), 15–22; and J. Garreau, *Edge City: Life on the New Frontier* (New York: Knopf Doubleday/Anchor Books, 1991).

27. A. C. Nelson, "The New Urbanity: The Rise of a New America," *Annals of the American Academy of Political and Social Science* 626 (2009): 192–208, 206.

28. Ibid., 207.

29. A. Flint, *This Land: The Battle over Sprawl and the Future of America* (Baltimore, MD: Johns Hopkins University Press, 2006), 59.

30. See J. K. De Jong, *New SubUrbanisms* (London: Routledge, 2014); and E. Talen, *New Urbanism and American Planning: The Conflict of Cultures* (London: Routledge, 2005), 1–2.

31. J. Kotkin, *The Next Hundred Million: America in 2050* (New York: Penguin Press, 2010), x.

32. N. A. Phelps, "The Growth Machine Stops? Urban Politics and the Making and Re-making of an Edge City," *Urban Affairs Review* 84 (2012): 670–700.

33. Harris, "Meaningful Types in a World of Suburbs."

34. See W. T. Bogart, *Don't Call It Sprawl: Metropolitan Structure in the 21st Century* (Cambridge: Cambridge University Press, 2006); and M. Orfield, *American Metropolitics: The New Suburban Reality* (Washington, DC: Brookings Institution Press, 2002).

35. R. Lang and P. Knox, "The New Metropolis: Rethinking Megalopolis," *Regional Studies* 43 (2009): 789–802.

36. R. Lang, *Edgeless Cities: Exploring the Elusive Metropolis* (Washington, DC: Brookings Institution Press, 2003); R. Fishman, *Bourgeois Utopias: The Rise and Fall of Suburbia* (New York: Basic Books, 1987); R. Kling, S. C. Olin, Jr., and M. Poster, "The Emergence of Postsuburbia: An Introduction," in *Postsuburban California: The*

Transformation of Orange County Since World War II, ed. R. Kling, S. Olin, and M. Poster (Berkeley: University of California Press, 1995), 1–30.

37. Teaford, *Post-Suburbia;* Beauregard, *When America Became Suburban.*

38. For one of the most precise definitions of post-suburbia in terms of a distinct era, see W. H. Lucy and D. L. Phillips, "The Postsuburban Era Comes to Richmond: City Decline, Suburban Transition and Exurban Growth," *Landscape and Urban Planning* 36 (1997): 259–275.

39. U. Beck, *Risk Society: Toward a New Modernity* (London: Sage, 1992).

40. Kling, Olin, and Poster, 'The Emergence of Postsuburbia."

41. S. Graham and S. Marvin, *Splintering Urbanism* (London: Routledge, 2000).

42. M. Dear, "The Los Angeles School of Urbanism: An Intellectual History," *Urban Geography* 24 (2003): 493–509.

43. Teaford, *Post-Suburbia.*

44. Collective consumption refers to use of a raft of services that, because of their public good properties or the large scale typically required for their production, have generally been provided by the state.

45. Lang, *Edgeless Cities.* Robert Lang's figures for 1999 suggest that as much as one-third of all office space in thirteen metropolitan areas of the United States was in what he terms edgeless cities—nearly twice that in edge cities. See also B. Scheer and M. Petkov, "Edge City Morphology: A Comparison of Commercial Centers," *Journal of the American Planning Association* 64 (1998): 298–310.

46. J. Nijman and T. Clery, "Rethinking Suburbia: A Case Study of Metropolitan Miami," *Environment and Planning A* 46 (2015): 69–88.

47. R. A. Mohl and G. Mormino, "The Big Change in the Sunshine State: A Social History of Modern Florida," in *The New History of Florida,* ed. M. Gannon (Miami: University Press of Florida, 1996), 418–447, 436.

48. Garreau, *Edge City.*

49. R. Bruegmann, *Sprawl: A Compact History* (Chicago: University of Chicago Press, 2005).

Chapter 2: Locating Post-Suburbs in a Metropolitan Context

1. See W. T. Bogart, *Don't Call It Sprawl: Metropolitan Structure in the Twenty-First Century* (Cambridge: Cambridge University Press, 2006); and M. Orfield, *American Metropolitics: The New Suburban Reality* (Washington, DC: Brookings Institution Press, 2002).

2. See R. Lang and P. Knox, "The New Metropolis: Rethinking Megalopolis," *Regional Studies* 43 (2009): 789–802; and J. Nijman and T. Clery, "Rethinking Suburbia: A Case Study of Metropolitan Miami," *Environment and Planning A* 46 (2015): 69–88.

3. B. Hanlon, T. Vicino, and J. R. Short, "The New Metropolitan Reality in the US: Rethinking the Traditional Model," *Urban Studies* 43 (2006): 2129–2143; N. Smith "Gentrification and Uneven Development," *Economic Geography* 58 (1982): 138–155.

4. See M. Castells, *The Urban Question,* trans. Alan Sheridan (Cambridge, MA: MIT Press, 1977); and P. Saunders, *Social Theory and the Urban Question* (London: Hutchison, 1981).

5. N. Brenner, D. J. Madden, and D. Wachsmuth, "Assemblages, Actor-networks, and the Challenges of Critical Urban Theory," in *Cities for People, Not for Profit: Critical Urban Theory and the Right to the City,* ed. N. Brenner, P. Marcuse, and M. Mayer (London: Routledge, 2011), 117–137; N. Brenner, "What Is Critical Urban Theory?," in Brenner, Marcuse, and Mayer, *Cities for People, Not for Profit,* 11–23.

6. R. Harris and P. J. Larkham, "Suburban Foundation, Form and Function," in *Changing Suburbs: Foundation, Form and Function,* ed. R. Harris and P. J. Larkham (London: Spon, 1999), 1–31.

7. R. Harris, "Meaningful Types in a World of Suburbs," in *Suburbanization in Global Society,* ed. M. Clapson and R. Hutchinson (Bingley, UK: Emerald, 2010), 15–50.

8. N. A. Phelps, A. Tarazona Vento, and S. Roitman, "The Suburban Question: Grassroots Politics and Place Making in Spanish Suburbs," *Environment and Planning C* 33 (3): 512–532.

9. A. Forsyth, "Defining Suburbs," *Journal of Planning Literature* 27 (2012): 270–281; Harris, "Meaningful Types in a World of Suburbs."

10. Harris and Larkham, "Suburban Foundation, Form and Function," 8.

11. Harris, "Meaningful Types in a World of Suburbs."

12. R. E. Lang, *Edgeless Cities: Exploring the Elusive Metropolis* (Washington, DC: Brookings Institution Press, 2002).

13. G. Hise, *Magnetic Los Angeles: Planning the Twentieth-Century Metropolis* (Baltimore, MD: Johns Hopkins University Press, 1999), 12.

14. R. Bruegmann, *Sprawl: A Compact History* (Chicago: University of Chicago Press, 2005), 71.

15. For my purposes I use the terms "reflexive" or "second" modernity, which place greater stress on the continuities apparent between processes and patterns of suburbanization and post-suburbanization. The terminological preference is important as it has implications for incorporating an understanding of the legacies of state interventions and their side effects in the production of suburbs and post-suburbs.

16. D. Hayden, *Building Suburbia: Green Fields and Urban Growth, 1820–2000* (New York: Vintage, 2004), 158.

17. J. C. O'Connell, *The Hub's Metropolis: Greater Boston's Development from Railroad Suburbs to Smart Growth* (Cambridge, MA: MIT Press, 2013).

18. R. Walker and R. D. Lewis, "Beyond the Crabgrass Frontier: Industry and the Spread of North American Cities, 1850–1950," *Journal of Historical Geography* 27 (2001): 3–19, 8–9.

19. R. A. Beauregard, *When America Became Suburban* (Minneapolis: University of Minnesota Press, 2006), 37.

20. R. McManus and P. J. Ethington, "Suburbs in Transition: New Approaches to Suburban History," *Urban History* 34 (2007): 317–337, 317.

21. Ibid., 325.

22. W. H. Lucy and D. L. Phillips, "The Postsuburban Era comes to Richmond: City Decline, Suburban Transition and Exurban Growth," *Landscape and Urban Planning* 36 (1997): 259–275, 261.

23. J. Garreau, *Edge City: Life on the New Frontier* (New York: Knopf Doubleday/ Anchor Books, 1991).

24. R. Fishman, *Bourgeois Utopias: The Rise and Fall of Suburbia* (New York: Basic Books, 1989), 17.

25. Garreau, *Edge City*. An edge city is defined as having at least 2 million square feet of office space, 600,000 square feet of retail space, more jobs than bedrooms, and is identified as a destination but was nothing like this thirty years ago.

26. See, among others, M. Dear, "The Los Angeles School of Urbanism: An Intellectual History," *Urban Geography* 24 (2003): 493–509; M. Dear and S. Flusty, "Postmodern Urbanism," *Annals of the Association of American Geographers* 88 (1998): 50–72; M. J. Dear, H. E. Schockman, and G. Hise, eds., *Rethinking Los Angeles* (Thousand Oaks, CA: Sage, 1996); and E. Soja, *Postmetropolis: Critical Studies of Cities and Regions* (Oxford: Wiley-Blackwell, 2000).

27. M. J. Dear and N. Dahmann, "Urban Politics and the Los Angeles School of Urbanism," *Urban Affairs Review* 44 (2008): 266–279, 269.

28. S. J. Essex and G. P. Brown, "The Emergence of Post-Suburban Landscapes on the North Coast of New South Wales: A Case Study of Contested Space," *Interna-*

tional Journal of Urban and Regional Research 21 (1997): 259–285; Lucy and Phillips, "The Postsuburban Era Comes to Richmond."

29. Lucy and Phillips, "The Postsuburban Era Comes to Richmond," 260 and 259, respectively.

30. N. Brenner, "Decoding the Newest 'Metropolitan Regionalism' in the USA: A Critical Overview," *Cities* 19 (2002): 3–21, 11.

31. N. A. Phelps, N. Parsons, D. Ballas, and A. Dowling, *Post-Suburban Europe: Planning and Politics at the Margins of Europe's Capital Cities* (Basingstoke: Palgrave Macmillan, 2006).

32. Bogart, *Don't Call It Sprawl*, 6–7.

33. Beauregard, *When America Became Suburban*, 65.

34. J. W. R. Whitehand and C. M. H. Carr, *Twentieth-Century Suburbs: A Morphological Approach* (London: Routledge, 2001), 121.

35. P. Larkham, "The Study of Urban Form in Great Britain," *Urban Morphology* 10 (2006): 117–150, 127. Urban morphological approaches to suburbanization have arrived at conclusions similar to those found in the literature on post-suburbia and postmodern urban theory primarily because the complexity of suburban morphology cannot be captured in terms such as innovation, diffusion, and distance decay, which themselves speak to a predictable linear pattern of evolution centered on the city.

36. H. W. Dick and P. J. Rimmer, "Beyond the Third World City: The New Urban Geography of South-east Asia," *Urban Studies* 35 (1998): 2303–2321.

37. See H. Nuissl and D. Rink, "The 'production' of Urban Sprawl in Eastern Germany as a Phenomenon of Post-Socialist Transition," *Cities* 22 (2005): 123–134; and M. Bontje and J. Burdack, "Edge Cities European-style: Examples from Paris and the Randstad," *Cities* 22 (2005): 317–330, 317.

38. K. T. Jackson, *Crabgrass Frontier: The Suburbanization of the United States* (Oxford: Oxford University Press, 1985), 303.

39. M. Gottdiener and G. Kephart, "The Multinucleated Region: A Comparative Analysis," in *Postsuburban California: The Transformation of Orange County since World War II*, ed. R. Kling, S. C. Olin, Jr., and M. Poster (Berkeley: University of California Press, 1995), 31–54, 51.

40. See Dear and Dahmann, "Urban Politics and the Los Angeles School of Urbanism"; R. Keil, "Global Sprawl: Urban Form after Fordism, *Environment and Planning D* 12 (1994): 31–36; and Soja, *Postmetropolis*. However, discussion of a distinctive geography of post-suburbia is also clouded by some of the successive simplifications made regarding the spatial structure of cities and suburbs associated with the spread

of the Chicago school ideas. See R. Harris and R. Lewis, "Constructing a Fault(y) Zone: Misrepresentations of American Cities and Suburbs, 1900–1950," *Annals of the Association of American Geographers* 88 (1998): 622–639.

41. Dear and Dahmann, "Urban Politics and the Los Angeles School of Urbanism," 269.

42. K. Brake et al., cited in C. Kraemer, "Commuter Belt Turbulence in a Dynamic Region: The Case of the Munich City-Region," in *The City's Hinterland: Dynamism and Divergence in Europe's Peri-Urban Territories,* ed. K. Hoggart (Aldershot: Ashgate, 2005), 41–68, 44.

43. McManus and Ethington, "Suburbs in Transition," 332.

44. R. Koolhaas, "Atlanta," in *Shaping the City: Studies in History, Theory, and Urban Design,* ed. R. El-Khoury and E. Robbins (London: Routledge, 2003), 5–13quoted in S. Marshall, "The Emerging 'Silicon Savanna': From Old Urbanism to New Suburbanism," *Built Environment* 32 (2006): 267–280, 268.

45. Fishman, *Bourgeois Utopias,* 203.

46. R. Fishman, "The Garden City Tradition in the Post-Suburban Age," *Built Environment* 17 (1991): 232–241, 234–235.

47. Lang, *Edgeless Cities,* 1–2.

48. S. Graham and S. Marvin, *Splintering Urbanism* (London: Routledge, 2001).

49. J. Gottmann, *Megalopolis* (New York: Twentieth Century Fund, 1961); T. G. McGee, "The Emergence of Desakota Regions in Asia: Expanding a Hypothesis," in *The Extended Metropolis: Settlement Transition in Asia,* ed. N. S. Ginsburg, B. Koppel, and T. G. McGee (Honolulu: University of Hawaii Press, 1991), 3–25.

50. Fishman, "The Garden City Tradition in the Post-Suburban Age," 239.

51. R. Kling, S. C. Olin, Jr., and M. Poster, "The Emergence of Postsuburbia: An Introduction," in *Postsuburban California: The Transformation of Orange County since World War II,* ed. R. Kling, S. Olin, and M. Poster (Berkeley: University of California Press, 1995), 1–30, 7.; Thomas Sieverts, *Cities without Cities: An Interpretatio of the Zwischenstadt* (London: Routledge, 2003).

52. Kraemer, "Commuter Belt Turbulence in a Dynamic Region," 44.

53. Gottdiener and Kephart, "The Multinucleated Region," and Dear and Flusty, "Postmodern Urbanism."

54. Dear and Dahmann, "Urban Politics and the Los Angeles School of Urbanism," 270.

55. Larkham, "The Study of Urban Form in Great Britain," 126–127. The fringe belt he describes as "a zone of largely extensive land uses that is formed at the edges of

an urban area during a pause in outward residential growth. Each fringe belt . . . has several distinctive features in terms of plan, building form, and land and building uses. Typical uses requiring extensive sites, include public utilities, parks, sports facilities, and allotment gardens."

56. Whitehand and Carr, *Twentieth-Century Suburbs.*

57. S. M. Wheeler, "The Evolution of Built Landscapes in Metropolitan Regions," *Journal of Planning Education and Research* 27 (2008): 400–416.

58. F. Wu and N. A. Phelps, "(Post) Suburban Development and State Entrepreneurialism in Beijing's Outer Suburbs," *Environment and Planning A* 43 (2011): 410–430. In major city-regions such as Beijing, the gentrification of urban cores occurs alongside the development of new mass residential suburbs and at the same time as extremely low-density ex-urban development.

59. Phelps et al., *Post-Suburban Europe.*

60. E. Mazierska and L. Rascaroli, *From Moscow to Madrid: Postmodern Cities, European Cinema* (London: I. B. Taurus, 2003), 18.

61. Although regulation theory has been developed in ways that speak to the local modes of regulation, it has rarely been used at the urban scale. In theories of the growth machine and urban regimes, regions and cities are treated as undifferentiated wholes.

62. M. Feldman, "Spatial Structures of Regulation and Urban Regimes," in *Reconstructing Urban Regime Theory: Regulating Urban Politics in a Global Economy,* ed. M. Lauria (London: Sage, 1997), 30–50, 31.

63. H. Molotch and J. Logan, "Tensions in the Growth Machine: Overcoming Resistance to Value-Free Development," *Social Problems* 31 (1984): 483–499.

64. Hise, *Magnetic Los Angeles.*

65. G. Frug, *City Making: Building Communities without Building Walls* (Princeton, NJ: Princeton University Press, 2001), 108.

66. D. Young and R. Keil, "Reconnecting the Disconnected: The Politics of Infrastructure in the In-Between City," *Cities* 27 (2010): 87–95, 94 and 93, respectively.

67. A. Althubaity and A. E. G. Jonas, "Suburban Entrepreneurialism: Redevelopment Regimes and Co-ordinating Metropolitan Development in Southern California," in *The Entrepreneurial City: Geographies of Politics, Regime and Representation,* ed. T. Hall and P. Hubbard (Chichester: John Wiley & Sons, 1998), 149–172.

68. On the economics of the in-between economy, see N. A. Phelps, "Clusters, Dispersion and the Spaces in between: For an Economic Geography of the Banal," *Urban Studies* 41 (2004): 971–989. On the politics of the in-between city, see Young and Keil, "Reconnecting the Disconnected."

69. Young and Keil, "Reconnecting the Disconnected," 87.

70. Garreau, *Edge City.*

71. Frug, *City Making,* 106.

72. See also J. Allen and A. Cochrane, "Beyond the Territorial Fix: Regional Assemblages, Politics and Power," *Regional Studies* 41 (2007): 1161–1175, 1172.

73. Feldman, "Spatial Structures of Regulation and Urban Regimes." Other means of orchestrating flows can be distinguished. For a discussion of the role of seduction within the relational properties of power, see J. Allen, *Lost Geographies of Power* (Chichester: Wiley-Blackwell, 2003). For an emphasis on models, reciprocity, and learning, see J. Braithwaite and J. Drahos, *Global Business Regulation* (Cambridge: Cambridge University Press, 2000).

74. Phelps, "Clusters, Dispersion and the Spaces in between"; N. A. Phelps and T. Ozawa, "Contrasts in Agglomeration: Proto-industrial, Industrial and Postindustrial Forms Compared," *Progress in Human Geography* 27(2003): 583–604.

75. T. Swanstrom, "Beyond Economism: Urban Political Economy and the Postmodern Challenge," *Journal of Urban Affairs* 15 (1993): 55–78.

76. As emphasized by, among others, D. Harvey, *The Urbanization of Capital,* 2 vols. (Baltimore, MD: Johns Hopkins University Press, 1985); R. Walker, "A Theory of Suburbanization: Capitalism and the Construction of Urban Space in the United States," in *Urbanization and Urban Planning in Capitalist Societies,* ed. M. Dear and A. Scott (London: Methuen, 1981), 383–429; and Walker and Lewis, "Beyond the Crabgrass Frontier."

77. Lang and Knox, "The New Metropolis," 790.

78. In Europe too these developments find their expression in morphologically and functionally polycentric patterns of urban development. See, for example, P. Hall and K. Pain, *The Polycentric Metropolis: Learning from Mega-City Regions in Europe* (London: Earthscan, 2006); and S. Musterd, M. Bontje, and W. Ostendorf, 'The Changing Role of Old and New Urban Centers: The Case of the Amsterdam Region," *Urban Geography* 27 (2006): 360–387. They may appear as muted, distorted, European versions of the likes of edge cities. See M. Bontje and M. Burdack, "Edge Cities European-style: Examples from Paris and the Randstad," *Cities* 22 (2005): 317–330.

79. Bogart, *Don't Call It Sprawl;* Lang and Knox, "The New Metropolis"; R. Shearmur, R. Coffee, C. Dube, and R. Barbonne, "Intrametropolitan Employment Structure: Polycentricity, Scatteration, Dispersal and Chaos in Toronto, Montreal and Vancouver, 1996–2001," *Urban Studies* 44 (2007): 1713–1738.

80. U. Beck, *Risk Society: Toward a New Modernity* (London: Sage, 1992); U. Beck, W. Bonss, and C. Lau, 'The Theory of Reflexive Modernization: Problematic, Hypotheses and Research Program," *Theory, Culture & Society* 20 (2003): 1–33.

81. Jackson, *Crabgrass Frontier,* 4.

82. Harris and Larkham, "Suburban Foundation, Form and Function."

83. Musterd, Bontje, and Ostendorf, "The Changing Role of Old and New Urban Centers." Notions of the homogeneity and stability of residential suburbs continue to have salience in Europe, although there too changes in the distribution of employment and populations have also begun to signal an urbanization of the suburbs, while a single European labor market and former colonial relations continue to fuel migration flows that have selectively altered the complexion of residential suburbs in the larger cities.

84. Hanlon, Vicino, and Short, "The New Metropolitan Reality in the US"; Orfield, *American Metropolitics.*

85. Dear and Dahmann, "Urban Politics and the Los Angeles School of Urbanism"; McManus and Ethington, "Suburbs in Transition."

86. Suburbs have rarely been residentially or industrially monofunctional: Hise, *Magnetic Los Angeles;* Walker and Lewis, "Beyond the Crabgrass Frontier."

87. J. Teaford, *Post-Suburbia: Government and Politics in the Edge Cities* (Baltimore, MD: Johns Hopkins University Press, 1997), 44.

88. L. H. Masotti, 'Prologue: Suburbia Reconsidered—Myth and Counter-myth," in *The Urbanization of the Suburbs,* ed. L. H. Masotti and J. K. Hadden (Beverly Hills, CA: Sage, 1973) 15–22, 16–17.

89. Fishman, *Bourgeois Utopias,* 190; O'Connell, *The Hub's Metropolis,* 202.

90. M. P. O'Mara, *Cities of Knowledge: Cold War Science and the Search for the Next Silicon Valley* (Princeton, NJ: Princeton University Press, 2005); L. Mozingo, *Pastoral Capitalism: A History of Suburban Corporate Landscapes* (Cambridge, MA: MIT Press, 2011).

91. N. A. Phelps, "Suburbs for Nations? Some Interdisciplinary Connections on the Suburban Economy," *Cities* 27 (2010): 68–76.

92. E. Dunham-Jones and J. Williamson, *Retrofitting Suburbia: Urban Design Solutions for Redesigning Suburbs* (Chichester: John Wiley & Sons, 2009).

Chapter 3: The Suburbs and Their Contradictions

1. A. J. Scott and S. T. Roweis, "Urban Planning in Theory and Practice: A Reappraisal," *Environment and Planning A* 9 (1977): 1097–1119; D. Harvey, *The Urbanization of Capital,* 2 vols. (Baltimore, MD: Johns Hopkins University Press, 1985).

2. U. Beck, *Risk Society: Toward a New Modernity* (London: Sage, 1992).

3. M. Dear, "The Los Angeles School of Urbanism: An Intellectual History," *Urban Geography* 24 (2004): 493–509.

4. J. Teaford, *Post-Suburbia: Government and Politics in the Edge Cities* (Baltimore, MD: Johns Hopkins University Press, 1997).

5. J. Habermas, *Legitimation Crisis* (Cambridge: Polity Press, 1988); C. Offe, "The Theory of the Capitalist State and the Problem of Policy Formation," in *Stress and Contradiction in Modern Capitalism: Public Policy and the Theory of the State,* ed. L. N. Lindberg, R. Alford, C. Crouch, and C. Offe (Lexington, MA: D. C. Heath, 1975), 125–144; C. Offe, *Disorganized Capitalism: Contemporary Transformations of Work and Politics* (Cambridge, MA: MIT Press, 1985).

6. R. Walker, "A Theory of Suburbanization: Capitalism and the Construction of Urban Space in the United States," in *Urbanization and Urban Planning in Capitalist Societies,* ed. M. Dear and A. Scott (London: Methuen, 1981), 383–429; Harvey, *The Urbanization of Capital.*

7. E. Talen, *New Urbanism and American Planning: The Conflict of Cultures* (London: Routledge, 2005), 13.

8. A. Wildavsky, "If Planning Is Everything, Maybe It's Nothing," *Policy Sciences* 4 (1973): 127–153.

9. Ibid.; E. Reade, "If Planning Is Anything, Maybe It Can Be Identified," *Urban Studies* 20 (1983): 159–171.

10. R. Bruegmann, *Sprawl: A Compact History* (Chicago: University of Chicago Press, 2005), 100.

11. Scott and Roweis, "Urban Planning in Theory and Practice," 1103.

12. Offe, "The Theory of the Capitalist State and the Problem of Policy Formation," 134.

13. M. H. Rose and R. A. Mohl, *Interstate: Highway Politics and Policy since 1939,* 3rd ed. (Knoxville: University of Tennessee Press, 2012).

14. C. E. Lindblom, "The Science of 'Muddling Through,'" *Public Administration Review* 19 (1959): 79–88.

15. W. T. Bogart, *Don't Call it Sprawl: Metropolitan Structure in the Twenty-First Century* (Cambridge: Cambridge University Press, 2006).

16. P. Hall, *Great Planning Disasters* (London: Weidenfeld, 1980).

17. Scott and Roweis, "Urban Planning in Theory and Practice," 1110.

18. Harvey, *The Urbanization of Capital,* 60–61.

19. Offe, "The Theory of the Capitalist State and the Problem of Policy Formation," 135.

20. J. C. Scott, *Seeing Like a State* (New Haven, CT: Yale University Press, 2000).

21. A. Harding, "Urban Regimes and Growth Machines: Toward a Cross-National Research Agenda," *Urban Affairs Quarterly* 29 (1994): 356–382; A. Harding, "Urban Regimes in a Europe of the Cities?," *European Urban and Regional Studies* 4 (1997): 291–314.

22. J. C. O'Connell, *The Hub's Metropolis: Greater Boston's Development from Railroad Suburbs to Smart Growth* (Cambridge, MA: MIT Press, 2013), 135.

23. Walker, "A Theory of Suburbanization," 409.

24. A. Flint, *This Land: The Battle over Sprawl and the Future of America* (Baltimore, MD: Johns Hopkins University Press, 2006); R. Hogan, *The Failure of Planning: Permitting Sprawl in San Diego Suburbs, 1970–1999* (Columbus: Ohio State University Press, 2003); A. Rome, *The Bulldozer in the Countryside: Suburban Sprawl and the Rise of American Environmentalism* (Cambridge: Cambridge University Press, 2001).

25. A. Duany, E. Plater-Zyberk, and J. Speck, *Suburban Nation: The Rise of Sprawl and the Decline of the American Dream* (New York: North Point Press, 2000), 4.

26. M. Gottdiener, *Planned Sprawl: Private and Public Interests in Suburbia* (Beverly Hills, CA: Sage, 1977).

27. Harvey, *The Urbanization of Capital,* 122.

28. Beck, *Risk Society,* 2.

29. R. Keil, "Global Sprawl: Urban Form after Fordism," *Environment and Planning D,* 12 (1994): 31–36.

30. D. Young and R. Keil, "Reconnecting the Disconnected: The Politics of Infrastructure in the In-Between City," *Cities* 27 (2010): 87–95, 90.

31. Offe, "The Theory of the Capitalist State and the Problem of Policy Formation."

32. Habermas, *Legitimation Crisis,* 72.

33. Beck, *Risk Society,* 47.

34. U. Beck, W. Bonss, and C. Lau, "The Theory of Reflexive Modernization: Problematic, Hypotheses and Research Program," *Theory, Culture & Society,* 20 (2003): 1–33.

35. Becker and Jahn, quoted in R. Keil, "Sustaining Modernity, Modernizing Nature: The Environmental Crisis and the Survival of Capitalism," in *The Sustainable Development Paradox: Urban Political Economy in the United States,* ed. R. Krueger and D. Gibbs (New York: Guilford Press, 2007), 41–65, 54.

36. On the application of ideas of communicative rationality in collaborative and deliberative planning, see P. Healey, *Collaborative Planning: Shaping Places in Fragmented Societies* (Basingstoke: Macmillan, 1997); J. Forester, *Planning in the Face of Power* (Berkeley: University of California Press, 1989). On the overburdening of the state, see Offe, "The Theory of the Capitalist State and the Problem of Policy Formation," 140, and idem, *Disorganized Capitalism,* 313.

37. Gottdiener, *Planned Sprawl,* 67.

38. M. Sheller and J. Urry, "The City and the Car," *International Journal of Urban and Regional Research* 24 (2000): 737–757.

39. Duany, Plater-Zyberk, and Speck, *Suburban Nation,* 19.

40. Network externalities exist where the utility of consumers and the advantages to producers depend on the number of consumers using a product or service over time. M. L. and C. Shapiro, "Network Externalities, Competition and Compatibility," *American Economic Review* 75 (1985): 424–440; M. L. Katz and C. Shapiro, "Technology Adoption in the Presence of Network Externalities," *Journal of Political Economy* 94 (1986): 822–841.

41. Sheller and Urry, "The City and the Car."

42. L. Lerup, "American Speed, American Distance," in *New Urbanism: Peter Calthorpe vs. Lars Lerup.,* ed. R. Fishman, Michigan Debates on Urbanism (Ann Arbor: University of Michigan, Taubman College of Architecture, 2005), 40–52, 43.

43. Robert A. Beauregard, *When America Became Suburban* (Minneapolis: University of Minnesota Press, 2006).

44. L. H. Masotti, "Prologue: Suburbia Reconsidered—Myth and Counter-myth," in *The Urbanization of the Suburbs,* ed. L. H. Masotti and J. K. Hadden (Beverly Hills, CA: Sage, 1973), 115–122; E. Dunham-Jones and J. Williamson, *Retrofitting Suburbia: Urban Design Solutions for Redesigning Suburbs* (Chichester: John Wiley & Sons, 2009). G. Tachieva, *Sprawl Repair Manual* (Washington, DC: Island Press, 2010).

45. Dunham-Jones and Williamson, *Retrofitting Suburbia,* v.

46. Bruegmann, *Sprawl,* 64.

47. J. Kotkin, *The Next Hundred Million: America in 2050* (New York: Penguin, 2010); A. C. Nelson, 'The New Urbanity: The Rise of a New America," *Annals of the American Academy of Political and Social Science* 626 (2009): 192–208.

48. Kotkin, *The Next Hundred Million,* 16.

49. A. Passell, *Building the New Urbanism: Places, Professions, and Profits in the American Metropolitan Landscape* (London: Routledge, 2013), 15.

50. Ibid., 2.

51. Dear, 'The Los Angeles School of Urbanism," 503.

52. Passell, *Building the New Urbanism,* 34.

53. Dunham-Jones and Williamson, *Retrofitting Suburbia,* 177.

54. T. Swanstrom, "Beyond Economism: Urban Political Economy and the Post-modern Challenge," *Journal of Urban Affairs* 15 (1993): 55–78.

55. Harding, "Urban Regimes and Growth Machine," 359.

56. Swanstrom, "Beyond Economism."

57. Harding, "Urban Regimes and Growth Machines."

58. M. Aglietta, *A Theory of Capitalist Regulation: The U.S. Experience* (London: Verso, 1979); J. O'Connor, *The Fiscal Crisis of the State* (New York: St. Martin's Press, 1973); J. O'Connor, *Accumulation Crisis* (Oxford: Basil Blackwell, 1984).

59. M. Goodwin and J. Painter, "Concrete Research, Urban Regimes and Regulation Theory," in *Reconstructing Urban Regime Theory: Regulating Urban Politics in a Global Economy,* ed. M. Lauria (London: Sage, 1997), 13–29; M. Feldman, "Spatial Structures of Regulation and Urban Regimes, in Lauria, *Reconstructing Urban Regime Theory,* 30–50; K. Ward, "Rereading Urban Regime Theory: A Sympathetic Critique," *Geoforum* 27 (1996): 427–438.

60. H. Molotch, "The City as a Growth Machine: Toward a Political Economy of Place," *American Journal of Sociology* 82 (1976): 309–32.

61. Ibid., 331.

62. C. Webster, "Property Rights and the Public Realm: Gates, Greenbelts, and Gemeinschaft," *Environment and Planning B* 29 (2002): 397–412, 409.

63. C. Stone, *Regime Politics: Governing Atlanta, 1946–1988* (Lawrence: University Press of Kansas, 1989), 8.

64. Ibid., 221.

65. J. Teaford, *Post-Suburbia: Government and Politics in the Edge Cities* (Baltimore, MD: Johns Hopkins University Press, 1997).

66. G. Frug, *City Making: Building Communities without Building Walls* (Princeton, NJ: Princeton University Press, 2001), 83.

67. K. T. Jackson, "Introduction: The Shape of Things to Come: Urban Growth in the South and West," in *Essays on Sunbelt Cities and Recent Urban America,* ed. R. Mohl, R. Fisher, C. Abbott, R. W. Lotchin, R. B. Fairbanks, and Z. L. Miller (College Station: Texas A&M University Press, 1990), 3–7, 5.

68. N. A. Phelps, N. Parsons, D. Ballas, and A. Dowling, *Post-Suburban Europe: Planning and Politics at the Margins of Europe's Capital Cities* (Basingstoke: Palgrave Macmillan, 2006); Teaford, *Post-Suburbia.*

69. Teaford, *Post-Suburbia.*

70. J. Logan and H. Molotch, *Urban Fortunes: The Political Economy of Place* (Berkeley: University of California Press, 1987), xx.

71. N. Smith, "Gentrification and Uneven Development," *Economic Geography* 58 (1982): 138–155.

72. F. J. Turner, *The Frontier in American History* (New York: Holt, Rinehart and Winston, 1962 [1920]).

73. J. Garreau, *Edge City: Life on the New Frontier* (New York: Knopf Doubleday/Anchor Books, 1991).

74. Compare J. M. Levy, *Contemporary Urban Planning* (Englewood Cliffs, NJ: Prentice Hall, 1988), ix, and P. McAuslan, *The Ideologies of Planning Law* (Oxford: Pergamon Press, 1980), on the legal basis of planning in the United States and the United Kingdom.

75. Molotch, "The City as a Growth Machine."

76. Jackson, "Introduction: The Shape of Things to Come," 4.

77. A. Sarzynski, G. Galster, and L. Stack, "Evolving United States Metropolitan Land Use Patterns," *Urban Geography* 35 (2014): 25–47, 25 and 26, respectively. Using a range of indicators of metropolitan land use, the authors typify patterns of urban development in the United States in terms of "intensity," "compactness," "mixing," and "core-dominance."

78. Z. L. Miller, (1990) "Pluralizing America: Walter Prescott Webb, Chicago School Sociology and Cultural Regionalism," in Mohl et al., *Essays on Sunbelt Cities and Recent Urban America,* 151–176; Talen, *New Urbanism and American Planning,* 1–2.

79. L. S. Bourne, "On Schools of Thought, Comparative Research, and Inclusiveness: A Commentary," *Urban Geography* 29 (2008): 177–186, 179.

80. J. H. Mollenkopf, "School Is Out: The Case of New York City," in *The City Revisited: Urban Theory from Chicago, Los Angeles, and New York,* ed. D. R. Judd and D. Simpson (Minneapolis: University of Minnesota Press, 2011), 169–185, 175, emphasis added.

81. R. E. Lang and J. B. LeFurgy, *Boomburbs: The Rise of America's Accidental Cities* (Washington, DC: Brookings Institution Press, 2007).

82. T. S. Chapin, "From Growth Controls to Comprehensive Planning to Smart Growth: Planning's Emerging Fourth Wave," *Journal of the American Planning Association* 78 (2012): 5–15, 6.

83. C. Abbott, "Southwestern Cityscapes: Approaches to an American Urban Environment," in Mohl et al., *Essays on Sunbelt Cities and Recent Urban America,* 59–86.

84. Chapin, "From Growth Controls to Comprehensive Planning to Smart Growth." Chapin identifies four eras of urban growth management in the United States: the era of growth controls, which did indeed involve a measure of containment through the use of the likes of stop lines and urban growth boundaries; the era of comprehensive planning, which did not seek to limit the spatial extent of urban development but instead tied it to questions of financing and concurrent provision for collective consumption; the era of smart growth management, in which a greater mix and density of development were encouraged selectively in urban and suburban centers; and an era of sustainable growth management, which he sees emerging presently to encompass yet broader and longer-term concerns such as climate change and food security alongside those of the resilience and sustainability of communities in economic development terms.

85. Teaford, *Post-Suburbia.*

86. Talen, *New Urbanism and American Planning,* 4.

87. A. Forsyth, "Who Built Irvine? Private Planning and the Federal Government," *Urban Studies* 39 (2002): 2507–2530.

88. R. E. Lang, *Edgeless Cities: Exploring the Elusive Metropolis* (Washington, DC: Brookings Institution Press, 2003).

89. Lang and LeFurgy, *Boomburbs.*

90. C. R. Leinberger, "The Next Slum?," *The Atlantic* (March 2008), 70–75, cited in Young and Keil, "Reconnecting the Disconnected," 92

Chapter 4: Politics and the Private and Collective Dynamics of America's Post-Suburban Future

1. R. Keil, "Sustaining Modernity, Modernizing Nature: The Environmental Crisis and the Survival of Capitalism," in *The Sustainable Development Paradox: Urban Political Economy in the United States,* ed. R. Krueger and D. Gibbs (New York: Guilford Press, 2007), 41–65, 62.

2. R. Gottlieb, *Reinventing Los Angeles: Nature and Community in the Global City* (Cambridge, MA: MIT Press, 2007), 332.

3. D. Hayden, *Building Suburbia: Greenfields and Urban Growth, 1820–2000* (New York, 2003), 3.

4. R. Bruegmann, *Sprawl: A Compact History* (Chicago: University of Chicago Press, 2005).

5. R. Fishman, *Bourgeois Utopias: The Rise and Fall of Suburbia* (New York: Basic Books, 1987).

6. Hayden, *Building Suburbia,* 245.

7. J. Teaford, *Post-Suburbia: Government and Politics in the Edge Cities* (Baltimore, MD: Johns Hopkins University Press1997), 5.

8. R. Kling, S. C. Olin, Jr., and M. Poster, "The Emergence of Postsuburbia: An Introduction," in *Postsuburban California: The Transformation of Orange County Since World War II,* ed. R. King, S. Olin, and M. Poster (Berkeley: University of California Press, 1995), 1–30.

9. M. J. Dear and N. Dahmann, "Urban Politics and the Los Angeles School of Urbanism," *Urban Affairs Review* 44 (2008): 266–279.

10. Bruegmann, *Sprawl,* 163.

11. D. Harvey, *The Urbanization of Capital,* 2 vols. (Baltimore, MD: Johns Hopkins University Press, 1985); R. Walker, "A Theory of Suburbanization: Capitalism and the Construction of Urban Space in the United States," in *Urbanization and Urban Planning in Capitalist Societies,* ed. M. Dear and A. Scott (London: Methuen, 1981), 383–429.

12. H. Molotch, "The City as a Growth Machine: Toward a Political Economy of Place," *American Journal of Sociology* 82 (1976): 309–322.

13. L. H. Masotti, "Prologue: Suburbia Reconsidered—Myth and Counter-myth," in *The Urbanization of the Suburbs,* ed. L. H. Masotti and J. K. Hadden (Beverly Hills, CA: Sage, 1973), 16–17.

14. L. Mozingo, *Pastoral Capitalism: A History of Suburban Corporate Landscapes* (Cambridge, MA: MIT Press, 2011), 217.

15. M. P. O'Mara, *Cities of Knowledge: Cold War Science and the Search for the Next Silicon Valley* (Princeton, NJ: Princeton University Press, 2005), 4.

16. R. Fishman, *Bourgeois Utopias: The Rise and Fall of Suburbia.* (New York: Basic Books, 1987), 190; R. Fishman, "The Garden City Tradition in the Postsuburban Age," *Built Environment* 17 (1991): 232–241, 234.

17. Teaford, *Post-Suburbia,* 44.

18. P. Saunders, *Social Theory and the Urban Question* (London: Hutchison, 1981); N. A. Phelps, N. Parsons, D. Ballas, and A. Dowling, *Post-Suburban Europe: Planning*

and Politics at the Margins of Europe's Capital Cities (Basingstoke: Palgrave Macmillan, 2006).

19. M. Gottdiener, *Planned Sprawl: Private and Public Interests in Suburbia* (Beverly Hills, CA: Sage, 1977).

20. P. O. Muller, "The Suburban Transformation of the Globalizing American City," *The Annals of the American Academy of Political and Social Science* 551 (1997): 44–58, cited in J. Kotkin, *The Next Hundred Million: America in 2050* (New York: Penguin, 2010), 94.

21. A. Wood, "The Scalar Transformation of the US Commercial Property-Development Industry: A Cautionary Note on the Limits of Globalization," *Economic Geography* 80 (2004): 119–140. On the "transnational capitalist class," see L. Sklair, "The Transnational Capitalist Class and Contemporary Architecture in Global Cities," *International Journal of Urban and Regional Research* 29 (2005): 484–500. On the idea of a "global intelligence corps," see K. Olds, *Globalization and Urban Change: Capital, Culture and Pacific Rim Mega-projects* (Oxford: Oxford University Press, 2001).

22. N. A. Phelps "Clusters, Dispersion and the Spaces in between: For an Economic Geography of the Banal," *Urban Studies* 41 (2004): 971–989.

23. T. Sieverts, *Cities without Cities: An Interpretation of the Zwischenstadt* (London: Routledge, 2003), 154.

24. R. A. Beauregard, *When America Became Suburban* (Minneapolis: University of Minnesota Press, 2006); Walker, "A Theory of Suburbanization"; J. Wolch, M. Pastor, and P. Dreier, "Introduction: Making Southern California: Public Policy, Markets, and the Dynamics of Growth," in *Up Against the Sprawl: Public Policy and the Making of Southern California*, ed. J. Wolch, M. Pastor, and P. Dreier (Minneapolis: University of Minnesota Press, 2004), 1–23. In the mixed economies of Europe, examples of the state instigating suburbanization and post-suburbanization are equally if not more pronounced, while in East Asia the state is currently perhaps *the* most important direct and indirect force driving suburban and post-suburban development in many national settings.

25. A. J. Scott and S. T. Roweis, "Urban Planning in Theory and Practice: A Reappraisal," *Environment and Planning A* 9 (1977): 1097–1119.

26. Gottdiener, *Planned Sprawl.*

27. U. Beck, W. Bonss, and C. Lau, "The Theory of Reflexive Modernization: Problematic, Hypotheses and Research Program," *Theory, Culture & Society* 20 (2003): 1–33.

28. E. Dunham-Jones and J. Williamson, *Retrofitting Suburbia: Urban Design Solutions for Redesigning Suburbs* (Chichester: John Wiley & Sons, 2009).

29. T. N. Clark, K. Lloyd, K. Wong, and P. Jain, "Amenities Drive Urban Growth," *Journal of Urban Affairs* 24 (2002): 493–515.

30. My reference to global here is not literal but is meant to imply simply a politics that exceeds purely local concerns centered on suburban communities, whether incorporated or not.

31. Scott and Roweis, "Urban Planning in Theory and Practice"; C. Offe, "The Theory of the Capitalist State and the Problem of Policy Formation," in *Stress and Contradiction in Modern Capitalism: Public Policy and the Theory of the State,* ed. L. N. Lindberg, R. Alford, C. Crouch, and C. Offe (Lexington, MA: D. C. Heath, 1975), 125–144.

32. M. Castells, *The Urban Question,* trans. Alan Sheridan (Cambridge, MA: MIT Press, 1977); idem, *The City and the Grassroots: A Cross-Cultural Theory of Urban Social Movements,* California Series in Urban Development 2 (Berkeley and Los Angeles: University of California Press, 1983).

33. R. Gottlieb, M. Vallianatos, R. M. Freeer, and P. Dreier, *The Next Los Angeles: The Struggle for a Livable City* (Berkeley: University of California Press, 2006); R. Gottlieb, *Reinventing Los Angeles: Nature and Community in the Global City* (Cambridge, MA: MIT Press, 2007).

34. R. E. Lang and J. B. LeFurgy, *Boomburbs: The Rise of America's Accidental Cities* (Washington, DC: Brookings Institution Press, 2007), 122.

35. Ibid., 121.

36. K. R. Cox and A. E. G. Jonas, "Urban Development, Collective Consumption and the Politics of Metropolitan Fragmentation," *Political Geography* 12 (1993): 8–37, 33.

37. J. Persky and W. Wiewel, *When Corporations Leave Town: The Costs and Benefits of Metropolitan Job Sprawl* (Detroit: Wayne State University Press, 2000), 82–83.

38. Beauregard, *When America Became Suburban.*

39. A. Passell, *Building the New Urbanism: Places, Professions, and Profits in the American Metropolitan Landscape* (London: Routledge, 2013), 18.

40. Beauregard, *When America Became Suburban,* 140.

41. A. While, A. E. G. Jonas, and D. C. Gibbs, "Unblocking the City? Growth Pressures, Collective Provision, and the Search for New Spaces of Governance in Greater Cambridge, England," *Environment and Planning A* 36 (2004) 279–304.

42. M. H. Rose and R. A. Mohl, *Interstate: Highway Politics and Policy since 1939,* 3rd ed. (Knoxville: University Press of Tennessee, 2012), 166. This much has become familiar in accounts of the suburban development of America, though it is worth

remembering that the interstate era actually emerged around a consensus regarding the presumed benefits of expressway development for central city regeneration, and not for the suburbs.

43. Ibid. Indeed, the politics surrounding urban expressway development was significantly imbued with a politics in which black and poor neighborhoods were demolished in order to facilitate fresh development opportunities for city business elites.

44. Persky and Wiewel, *When Corporations Leave Town.*

45. Rose and Mohl, *Interstate,* 188.

46. M. Parés and D. Saurí, "Integrating Sustainabilities in a Context of Economic, Social and Urban Change: The Case of Public Spaces in the Metropolitan Region of Barcelona," in Krueger and Gibbs, *The Sustainable Development Paradox,* 160–191, 187.

47. D. B. Massey, "A Global Sense of Place," *Marxism Today* 35 (1991): 24–29.

48. Kotkin, *The Next Hundred Million,* 215.

49. Harvey, *The Urbanization of Capital,* 122.

50. A. Duany, E. Plater-Zyberk, and J. Speck, *Suburban Nation: The Rise of Sprawl and the Decline of the American Dream* (New York: North Point Press, 2000), 42.

51. A. Rome, *The Bulldozer in the Countryside: Suburban Sprawl and the Rise of American Environmentalism* (Cambridge: Cambridge University Press, 2001), 229.

52. Dunham-Jones and Williamson, *Retrofitting Suburbia,* viii.

53. A. C. Nelson, "The New Urbanity: The Rise of a New America," *Annals of the American Academy of Political and Social Science* 626 (2009): 192–208, 206.

54. A. Flint, *This Land: The Battle over Sprawl and the Future of America* (Baltimore, MD: Johns Hopkins University Press, 2006), 103.

55. Dunham-Jones and Williamson, *Retrofitting Suburbia,* 177.

56. A. E. G. Jonas and S. Pincetl, "Rescaling Regions in the State: The New Regionalism in California," *Political Geography* 25 (2006): 482–505, 485.

57. R. Hogan, *The Failure of Planning: Permitting Sprawl in San Diego Suburbs, 1970–1999* (Columbus: Ohio State University Press, 2003) 87.

58. A. E. G. Jonas and A. While, "Greening the Entrepreneurial City? Looking for Spaces of Sustainability in the Competitive City," in Krueger and Gibbs, *The Sustainable Development Paradox,* 123–159, 144.

59. Gottdiener, *Planned Sprawl,* 25 and 111, respectively.

60. C. M. Tiebout, "A Pure Theory of Local Expenditures," *Journal of Political Economy* 64 (1956): 416–424.

61. Jonas and While, "Greening the Entrepreneurial City?," 146.

62. G. A. Gonzalez, *Urban Sprawl, Global Warming, and the Empire of Capital* (Albany: State University of New York Press, 2009).

63. R. Cervero and J. Murakami, "Effects of Built Environments on Vehicle Miles Traveled: Evidence from 370 Urbanized Areas," *Environment and Planning A* 42 (2010): 400–418.

64. T. S. Chapin, "From Growth Controls to Comprehensive Planning to Smart Growth: Planning's Emerging Fourth Wave," *Journal of the American Planning Association* 78 (2012): 5–15.

65. Gottlieb, *Reinventing Los Angeles.*

66. Ibid., 54.

67. Underbounding refers to a situation in which the functional economic area of a city or incorporated community (as measured in, for example, the travel to work area) exceeds the administrative boundary of the same community.

68. R. Mohl, "The Transformation of Urban America since the Second World War," in *Essays on Sunbelt Cities and Recent Urban America,* ed. R. Mohl, R. Fisher, C. Abbott, R. W. Lotchin, R. B. Fairbanks, and Z. L.Miller (College Station: Texas A&M University Press, 1990), 8–32, 24.

69. Teaford, *Post-Suburbia,* ix–x.

70. Ibid., 209–210.

71. Ibid., 84.

72. A. Forsyth, "Defining Suburbs," *Journal of Planning Literature* 27 (2012): 270–281.

73. Dunham-Jones and Williamson, *Retrofitting Suburbia,* 191.

74. Ibid., 179–180.

75. K. T. Jackson, *Crabgrass Frontier: The Suburbanization of the United States* (Oxford: Oxford University Press, 1985), 146.

76. Ibid., 140–141.

77. Ibid., 150.

78. Beauregard, *When America Became Suburban,* 189.

79. A. Silverman and L. Schneider, "Suburban Localism and Long Island's Regional Crisis," *Built Environment* 17 (1991): 191–204, 200–201.

80. Persky and Wiewel, *When Corporations Leave Town,* 99.

81. Ibid., 83.

82. M. Orfield, *American Metropolitics: The New Suburban Reality* (Washington, DC: Brookings Institution Press, 2002).

83. A. Althubaity and A. E. G. Jonas, "Suburban Entrepreneurialism: Redevelopment Regimes and Co-ordinating Metropolitan Development in Southern California," in *The Entrepreneurial City: Geographies of Politics, Regime and Representation,* ed. T. Hall and P. Hubbard (Chichester: John Wiley & Sons, 1998), 149–172, 149.

84. Ibid., 150.

85. J. A. Musso, "Metropolitan Fiscal Structure: Coping with Growth and Fiscal Constraint," in Wolch, Pastor, and Dreier, *Up Against the Sprawl,* 171–194.

86. Gottdiener, *Planned Sprawl,* 26.

87. Althubaity and Jonas, "Suburban Entrepreneurialism."

88. R. Keil, "Governance Restructuring in Los Angeles and Toronto: Amalgamation or Secession," *International Journal of Urban and Regional Research* 24 (2000): 758–781, 777.

89. N. Brenner, "Between Fixity and Motion: Accumulation, Territorial Organization, and the Historical Geography of Spatial Scales," *Environment and Planning D: Society and Space* 16 (1998): 459–481.

90. Chapin, "From Growth Controls to Comprehensive Planning to Smart Growth"; J. F. Horan and G. T. Taylor, *Experiments in Metropolitan Government* (New York: Praeger, 1977).

Chapter 5: Kendall Downtown

1. R. A. Mohl and G. Mormino, "The Big Change in the Sunshine State: A Social History of Modern Florida," in *The New History of Florida,* ed. R. A. Mohl and G. Mormino (Gainesville: University Press of Florida, 1996), 418–447, 425.

2. R. Gottlieb, M. Vallianatos, R. M. Freer, and P. Dreier, *The Next Los Angeles: The Struggle for a Livable City* (Berkeley: University of California Press, 2006).

3. See M. Grunwald, *The Swamp: The Everglades, Florida and the Politics of Paradise* (New York: Simon & Schuster, 2006); and C. Pittman and M. Waite, *Paving Paradise: Florida's Vanishing Wetlands and the Failure of No Net Loss* (Gainesville: University Press of Florida, 2009).

4. D. R. Colburn and L. deHaven-Smith, *Florida's Megatrends: Cultural Issues in Florida,* 2nd ed. (Gainesville: University Press of Florida, 2010), 135.

5. D. M. Parker, "Is South Florida the New Southern California? Carl Hiaasen's Dystopian Paradise," *Florida Historical Quarterly* 90 (2012): 306–323.

6. R. E. Lang, *Edgeless Cities: Exploring the Elusive Metropolis* (Washington, DC: Brookings Institution Press, 2003), 69.

7. J. Nijman, *Miami: Mistress of the Americas* (Philadelphia: University of Pennsylvania Press, 2010), 205.

8. G. R. Mormino, "Sunbelt Dreams and Altered States: A Social and Cultural History of Florida, 1950–2000," *Florida Historical Quarterly* 80 (2002): 3–21, 20.

9. L. Sandercock, *Cosmopolis II: Mongrel Cities of the 21st Century* (London: Bloomsbury Academic, 2003).

10. E. Dunham-Jones and J. Williamson, *Retrofitting Suburbia: Urban Design Solutions for Redesigning Suburbs* (Chichester: John Wiley & Sons, 2009); D. S. Goldstein, "New Urbanism: Recreating Florida by Rewriting the Rules," *Florida Bar Journal* 80 (2006): 63–71.

11. Mohl and Mormino, "The Big Change in the Sunshine State," 419.

12. Nijman, *Miami,* 67.

13. T. S. Chapin, "Growth Management or Growth Unabated? Economic Development in Florida since 1990," in *Growth Management in Florida: Planning for Paradise,* ed. T. S. Chapin, C. E. Connerly, and H. T. Higgins (Aldershot: Ashgate, 2007), 119–140, 137 and 139, respectively.

14. H. Molotch, "The City as a Growth Machine: Toward a Political-Economy of Place," *American Journal of Sociology* 82 (1976): 309–322.

15. Nijman, *Miami,* 70.

16. Ibid., 126.

17. P. S. George, "Kendall," in *Miami's Historic Neighborhoods: A History of Community,* ed. B. R. Matkov (San Antonio: Historical Publishing Network, 2001), 118–122, 122.

18. Ibid.

19. According to Dr. Stokes of Miami Dade Community College, South Campus, "One issue is that the area is not evidently self-contained. Residents must commute out of Kendall to their place of work. We have a multitude of skilled workers living in Kendall, but no industry to support their working here." Quoted in M. E. Pando, "Life in Fast-Growing Kendall," *Kendall Gazette* 18 (1992): 8B–9B.

20. George, "Kendall," 118.

21. C. M. Rogier, *The Comprehensive Development Master Plan: A Study of Environmental Politics* (Pittsburgh, PA: University of Pittsburgh, 1982), 37.

22. C. R. Leonard, *License to Steal: Secrets of Acquiring Distressed Property in Florida* (Jacksonville, FL: Cliff R. Leonard & Associates, 1982).

23. W. B. Stronge, *The Sunshine Economy: An Economic History of Florida since the Civil War* (Gainesville: University Press of Florida, 2008), 193.

24. A. Ettorre, *Arvida: A Business Odyssey* (Coral Springs, FL: ECI, 1990), 11–12.

25. Quoted in B. Weiner, "Dadeland Work Starts Tomorrow," *Miami News*, August 6, 1961, 7c.

26. P. Kaplan, "Kendall," in *Miami's Neighborhoods*, ed. P. Morrissey (Miami: Miami News, 1982), 64–66, 66.

27. Some measure of Arvida's interest in rapid infrastructure improvements in the Kendall area, represented by the likes of the North Kendall Drive extensions, is indicated by the fact that in 1958, at the announcement of the issue of its flotation on the U.S. stock exchange, only 1 percent of its land holdings had been improved. By as late as 1971 as little as 17 percent of Arvida's land holdings had been improved, but by the end of 1976, 68 percent of its land holdings had zoning, water and sewer service, and access roads, and were at elevations required for construction. Arvida Corporation, *Prospectus Class A Common Stock* (1956), and Arvida Corporation, *Annual Report* (1976), 2–3.

28. J. Taylor, *A History of South Dade County, Florida* (Miami, 1983), 5. One interviewee who arrived in the area in the late 1960s recalled how the road remained surrounded on either side by farmland west of the turnpike and was a prime spot for drag car racing at the time (Interview, planner, Metropolitan Division, Miami-Dade County Planning Department, June 26, 2013).

29. Interview, principal, Henry Wolffe Jr. and Associates, Miami Beach, June 26, 2013.

30. P. S. George, *A Journey through Time: A Pictorial History of South Dade* (Virginia Beach, VA: Walsworth, 1995), 165.

31. Dunham-Jones and Williamson, *Retrofitting Suburbia*.

32. Goldstein, "New Urbanism," 63 and 69, respectively.

33. See http://www.chambersouth.com/community/downtown-kendall (accessed May 31, 2014).

34. P. Whoriskey, "Vision of a Downtown Intrigues Kendall," *Miami Herald*, July 19, 1998, 1A.

35. Ibid.

36. Interview, membership development manager, Miami-Dade Chamber of Commerce (South), April 21, 2011.

37. See http://www.miami21.org (accessed December 14, 2014). A form-based zoning code is less concerned less with specifying set densities of development (floor-to-area ratios, or FARs) than with architectural form.

38. Interview, planner, Community Planning, Miami Dade County Planning Department, April 15, 2011.

39. Interview, partner, Dover, Kohl and Partners, Coral Gables, June 28, 2013.

40. Interview, former mayor, Cutler Bay, April 28, 2011.

41. Interview, executive vice president, The Green Companies, Dadeland Center, June 27, 2013.

42. Interview, membership development manager, Miami-Dade Chamber of Commerce (South), April 21, 2011.

43. Interview, president and CEO, Good Governance Initiative, Miami, April 22, 2011.

44. Interview, executive vice president, The Green Companies, Dadeland Center, June 27, 2013.

45. Interview, former mayor, Cutler Bay, April 28, 2011.

46. Interview, executive vice president, The Green Companies, Dadeland Center, June 27, 2013.

47. Chapin, "Growth Management or Growth Unabated?," 1.

48. T. S. Chapin, C. E. Connerly and H. T. Higgins, "Introduction," in Chapin, Connerly, and Higgins, *Growth Management in Florida,* 1–4, 2.

49. Chapin, "Growth Management or Growth Unabated?," 9.

50. T. G. Pelham, "A Historical Perspective for Evaluating Florida's Evolving Growth Management Process," in Chapin, Connerly, and Higgins, *Growth Management in Florida,* 7–19, 15.

51. Colburn and deHaven-Smith, *Florida's Megatrends,* 127.

52. M. E. Klass, "Florida Lawmakers Wipe out 30 Years of Growth Management-Law," *Tampa Bay Times,* May 7, 2011.

53. M. S. Douglas, *The Everglades: River of Grass* (Sarasota: Pineapple Press Inc., 1988 [1947]).

54. Quoted in R. S. Turner and M. S. Murray, "Managing Growth in a Climate of Urban Diversity: South Florida's Eastward Ho! Initiative," *Journal of Planning Education and Research* 20 (2001): 308–328, 316.

55. Turner and Murray, "Managing Growth in a Climate of Urban Diversity," 324.

56. R. A. Mohl, "Interstating Miami: Urban Expressways and the Changing American City," *Tequesta* 68 (2008): 5–40.

57. Turner and Murray, "Managing Growth in a Climate of Urban Diversity," 326.

58. Interview, president and CEO, Good Governance Initiative, Miami, April 22, 2011.

59. K. Ross, "Downtown for Kendall Delayed a Year," *Miami Herald,* April 13, 2001, 3B.

60. Interview, executive vice president, The Green Companies, Dadeland Center, June 27, 2013.

61. Interview, vice president of property management, Lennar Corporation, April 21, 2011.

62. Interview, architect, Duany Plater-Zyberk & Company, April 26, 2011.

63. Interview, professor of architecture, University of Miami, April 22, 2011.

64. Interview, architect, Duany Plater-Zyberk & Company, April 26, 2011.

65. Interview, architect, Duany Plater-Zyberk & Company, April 26, 2011.

66. A. C. Nelson, "The New Urbanity: The Rise of a New America," *Annals of the American Academy of Political and Social Science* 626 (2009): 192–208.

67. Interview, president, Continental Park Residents Board, and member, East Kendall Homeowners Association, April 18, 2011.

68. Colburn and deHaven-Smith, *Florida's Megatrends,* 57.

69. Ibid., 122.

70. L. J. Carter, *The Florida Experience: Land and Water Policy in a Growth State* (Baltimore, MD: Johns Hopkins University Press, 1974), 138, 139.

71. Ibid., 155, 156.

72. Rogier, *The Comprehensive Development Master Plan,* 46.

73. Ibid., 48.

74. Miami-Dade County, *Comprehensive Development Master Plan for Metropolitan Dade County, Florida* (Miami, 1975), 136.

75. Ibid., 101.

76. P. Kaplan, "West Kendall Growth Not As Easy As ABC," *Miami News*, December 7, 1978, 1A and 6A.

77. Environmental Protection Agency (EPA), *Growing for a Sustainable Future: Miami-Dade County Urban Development Boundary Assessment* (Washington, DC: EPA, 2012), 1 and 7, respectively.

78. F. W. Boal, "Urban Growth and Land Value Patterns: Government Influences," *Professional Geographer* 22 (1970): 79–82.

79. EPA, *Growing for a Sustainable Future,* 7. Nevertheless, Miami Dade Commissioners recently agreed to expand the area within the UDB by 521 acres while also rejecting recommendations by planners to provide greater protection for 3,600 acres of environmentally sensitive land. This was the first such enlargement since 2011, when 120 acres were included. P. Mazzei, "Miami Dade Commissioners Expand Urban Development Boundary," *Miami Herald,* October 3, 2013.

80. EPA, *Growing for a Sustainable Future,* 4 and 10, respectively.

81. Ibid., 23.

82. Miami-Dade County, *Comprehensive Development Master Plan* (Miami, 1988), 1–15.

83. Interview, planning manager, Miami Dade Expressway Authority, April 25, 2011.

84. Miami-Dade County, *Comprehensive Development Master Plan for Metropolitan Dade County Florida*, 118.

85. Miami-Dade County, *Comprehensive Development Master Plan (revision) for Miami Dade County* (Miami, 1979), 163.

86. Miami-Dade County, *Comprehensive Development Master Plan for Metropolitan Dade County Florida* (Miami, 1975), 3.

87. M. Lucoff, "Metro, Cities Outline Transit Zoning Rules," *Miami News,* September 29, 1978, 4A.

88. EPA, *Growing for a Sustainable Future,* 22.

89. Interview, regional planner, South Florida Regional Planning Commission, April 27, 2011.

90. Interview, executive vice president, The Green Companies, Dadeland Center, June 27, 2013.

91. Interview, president and CEO, Good Governance Initiative, Miami, April 22, 2011.

92. Interview, director of engineering, Miami Dade Expressway Authority, April 25, 2011.

93. Ibid.

94. Interview, resident and local historian, East Kendall, April 26, 2011.

95. R. P. Wolff, *Miami Metro: The Road to Urban Unity* (Coral Gables, FL, 1960), 82.

96. Ibid., 58–59.

97. E. Sofen, *The Miami Metropolitan Experiment* (Bloomington: Indiana University Press, 1963), 13.

98. Wolff, *Miami Metro,* 82.

99. Sofen, *The Miami Metropolitan Experiment.*

100. Colburn and deHaven-Smith, *Florida's Megatrends,* 69.

101. Sofen, *The Miami Metropolitan Experiment,* 11.

102. J. F. Horan and G. T. Taylor, *Experiments in Metropolitan Government* (New York, 1977), 99.

103. Miami Dade County, *Comprehensive Development Master Plan (revision) Miami Dade County*, 165.

104. R. K. Vogel and G. N. L. Stowers, "Miami: Minority Empowerment and Regime Change," in *Big City Politics in Transition,* ed. H. V. Savitch and J. C. Thomas (London: Sage, 1991), 115–131, 121.

105. Lucoff, "Metro, Cities Outline Transit Zoning Rules," 16A.

106. Interview, professor of architecture, University of Miami, April 22, 2011.

107. Interview, president, Continental Park Residents Board, and member, East Kendall Homeowners Association, April 18, 2011.

108. Interview, resident and local historian, East Kendall, April 26, 2011.

109. Interview, planner, Community Planning, Miami Dade County Planning Department, April 15, 2011.

110. Colburn and deHaven-Smith, *Florida's Megatrends,* 138–139.

111. Nijman, *Miami,* 74.

112. Ibid., 136.

113. Ibid., 211.

114. Interview, member, Pinecrest Village Council, April 27, 2011.

Chapter 6: Taming Tysons

1. Interview, partner, KGP Design Studio, Washington, D.C., May 21, 2008.

2. J. Garreau, *Edge City: Life on the New Frontier* (New York: Knopf Doubleday/ Anchor Books, 1991).

3. Maryland-National Capital Park and Planning Commission, *On Wedges and Corridors: A General Plan for the Maryland-Washington Regional District in Montgomery and Prince George's Counties* (Washington, DC: Maryland-National Capital Park and Planning Commission 1962).

4. J. Peck, "Neoliberal Suburbanism: Frontier Space," *Urban Geography* 32 (2011): 884–919, 908, argues that the centrifugal momentum of suburban development continues to drive a metropolitan transformation in the United States involving "the evasion and subversion of post-urban regulatory coordination, and its secessionist eclipse through the prosaic medium of suburbanized 'subgovernance,'" where subgovernance entails not only a "a locally scaled modality of suburban (self) rule" but also "a purposive bundle of secessionist, exclusionary and marketizing interventions."

5. J. Logan and H. Molotch, *Urban Fortunes: The Political Economy of Place* (Berkeley: University of California Press, 1987).

6. R. E. Lang, *Edgeless Cities: Exploring the Elusive Metropolis* (Washington, DC: Brookings Institution Press, 2003).

7. C. P. Stuntz and M. Stuntz, *This Was Tysons Corner, Virginia: Facts and Photos* (Fairfax, VA: Stuntz and Stuntz, 1982).

8. P. L. Knox, "The Restless Urban Landscape: Economic and Sociocultural Change and the Transformation of Metropolitan Washington, DC," *Annals of the Association of American Geographers* 81 (1991): 181–209.

9. P. E. Ceruzzi, *Internet Alley: High Technology in Tysons Corner, 1945–2005* (Cambridge, MA: MIT Press, 2008).

10. Til Hazel was a lawyer acting on behalf of the state and federal governments identifying land and buildings to make way for the Beltway. He went on to become a tenacious land speculator and developer himself as well as acting as legal adviser in planning applications for some of the earliest developers at Tysons Corner.

11. Interview, former county executive, Fairfax County, April 9, 2008.

12. S. Mastran, "The Evolution of Suburban Nucleations: Land Investment Activity in Fairfax County, Virginia," PhD diss., University of Maryland, 1988; S. Mastran, "Tysons Corner, Virginia: Planning for the Urban Retrofitting of a Suburban Edge City," *Real Estate Review* 39 (2010): 63–81, 72.

13. Ceruzzi, *Internet Alley*, and Garreau, *Edge City*, document Til Hazel's and Gerry Halpin's roles in the physical development of and wider promotion of economic development in northern Virginia.

14. A. J. Scott and S. T. Roweis, "Urban Planning in Theory and Practice: A Reappraisal," *Environment and Planning A* 9 (1977): 1097–1119.

15. K. T. Jackson, *Crabgrass Frontier: The Suburbanization of the United States* (New York: Oxford University Press, 1985), 267.

16. Interview, president, Northern Virginia Transportation Alliance, May 29, 2008.

17. H. Molotch, "The City as a Growth Machine: Toward a Political Economy of Place," *American Journal of Sociology* 82 (1976): 309–332.

18. Ceruzzi, *Internet Alley*, 55.

19. B. Katz, "Welcome to the 'Exit Ramp' Economy," May 13, 2001, http://www.brookings.edu/research/opinions/2001/05/13metropolitanpolicy-katz (accessed May 25, 2014).

20. J. Teaford, *Post-Suburbia: Government and Politics in the Edge Cities* (Baltimore, MD: Johns Hopkins University Press, 1997).

21. R. Banham, *The Fight for Fairfax: A Struggle for a Great American County* (Fairfax, VA: George Mason University Press, 2009).

22. T. S. Peters, *The Politics and Administration of Land Use Control: The Case of Fairfax County, Virginia* (Lexington, MA: D. C. Heath, 1974).

23. Interview, former Fairfax County supervisor, May 27, 2008.

24. Mastran, "Tysons Corner, Virginia," 67.

25. Ibid., 67.

26. Banham, *The Fight for Fairfax;* N. Netherton, D. Sweig, J. Artemel, P. Hickin, and P. Reed. *Fairfax County, Virginia: A History* (Fairfax, VA: Fairfax County Board of Supervisors, 1978).

27. By this time, the rapid growth of the county's population, from approximately 96,000 in 1950 to 455,000 in 1970, had produced a measure of popular support for and the election of a "growth control" board of supervisors (Peters, *The Politics and Administration of Land Use Control*, 5). Popular support for this agenda was short-lived as a result of continuing failures of the board to deal adequately with these concerns at the same time as restraints on development increased the share of the tax take coming from residents. See County of Fairfax, *Committee to Study the Means of Encouraging Industrial Development in Fairfax County* (Fairfax, VA, 1976).

28. Ibid.

29. Interview, vice chair, Fairfax County Planning Commission, April 23, 2010.

30. Ceruzzi, *Internet Alley*.

31. Interview, former Fairfax County executive, April 9, 2008.

32. Interviews, former Fairfax County supervisor, May 28, 2008, and former Fairfax County supervisor, May 27, 2008. A planning aspiration to create a more pedestrian-friendly, urban center with a greater mix of uses has been apparent since 1971. See County of Fairfax, *Tysons Corner Regional Center Study* (Fairfax, VA, 1971). This was reiterated in County of Fairfax, *Tysons Corner Area Study* (Fairfax, VA, 1977). There was also an independent visioning exercise in the 1990s by the Tysons Transportation Association, *The Future of Tysons Corner: A Vision* (Tysons Corner, VA, 1992). The most recent comprehensive plan was County of Fairfax, *Fairfax County Comprehensive Plan* (Fairfax, VA, 1994). It is this plan that has been revised as a result of the recommendations of the Tysons Corner Land Use Task Force.

33. Banham, *The Fight for Fairfax,* 179; Netherton et al., *Fairfax County, Virginia.*

34. Interview, vice president, EDAW Inc., Alexandria, April 3, 2008.

35. Interview, president, Vienna-Tysons Regional Chamber of Commerce, April 8, 2008.

36. Interview, president and CEO, Fairfax County Economic Development Authority, April 4, 2008.

37. Interview, lawyer, Holland & Knight LLP, April 22, 2010.

38. "Tysons Corner, 2.0: Retrofitting a Giant, 50 Years Later," *Washington Post,* April 23, 2010, A20.

39. Knox, "The Restless Urban Landscape,"191.

40. Interview, senior vice president, Development Services, West Group, May 28, 2008.

41. Interview, vice president, EDAW Inc., Alexandria, April 3, 2008.

42. The task force was charged with promoting more mixed uses, better facilitating transit-oriented development, enhancing pedestrian connections across the Tysons area, increasing the residential component of the land-use mix, improving the functionality of Tysons, and providing for amenities and aesthetics in Tysons. County of Fairfax, *The Comprehensive Plan for Fairfax County* (Amendment No. 2007–23) (Fairfax, VA, 2010), 4.

43. Interviews, president and CEO, Fairfax County Chamber of Commerce, April 8, 2008, and mayor, Town of Vienna, May 23, 2008, respectively.

44. Interview, former Fairfax County supervisor, May 28, 2008.

45. George Mason University Center for Regional Analysis, *Forecasts for Tysons Corner to 2050* (Arlington, VA, 2008). The highest of a series of forecasts of future growth at Tysons Corner anticipates the population growing to 85,900 and employment growing to 209,900 by 2050, representing more or less a doubling of employment and a fivefold increase in population from that in 2008.

46. A. Gardner, "U.S. Transportation Chief Backs Dulles Rail Project," *Washington Post*, January 8, 2009, B1.

47. See the web page http://www.fairfaxcounty.gov/news/2013/board-approves-largest-residential-project-tysons.htm (accessed May 15, 2014).

48. County of Fairfax, *The Comprehensive Plan for Fairfax County* (Amendment No. 2007–23), 6.

49. Interview, partner, Davis Carter Scott, May 20, 2008.

50. Interview, president and CEO, Fairfax County Chamber of Commerce, April 8, 2008.

51. Interview, committee chairman, McLean Citizens Association, May 29, 2008.

52. County of Fairfax, *Tyson's Corner Urban Design Charrette* (Fairfax, VA, 1976), 13.

53. Mastran, "Tysons Corner, Virginia," 72.

54. The Los Angeles River, once natural, is now almost entirely artificial. For much of its length it passes through an elaborate set of concrete channels engineered for flood control purposes. Los Angeles effectively turned its back on the river, but civic and environmental groups have campaigned successfully to restore some of the amenity and ecological value of the river. See B. Gumprecht, *The Los Angeles River: Its Life, Death and Possible Rebirth* (new edition) (Baltimore, MD: Johns Hopkins University Press, 2001).

55. Interviews, former Fairfax County supervisor, May 28, 2008; president, The Regency Residents Association, June 13, 2008.

56. J. O'Connell, "Fairfax County Plans 154 Acres of Urban Parks for Tysons," *Washington Post*, April 25, 2014; Fairfax County Parks Authority, *Tysons Park System Concept Plan (Draft)* (Fairfax VA, 2014), vii.

57. T. Jackman, "Fairfax Agrees Third New Large Development for Tysons 'Other' Green Space," *Washington Post*, April 22, 2013. A coalition of fifteen neighborhood groups led the Save Tysons' Last Forest campaign to successfully help protect proposed exit ramp and road developments from the Dulles Toll Road into Tysons Corner from destroying an area of forest. See the website http://www.savetysonslastforest.org (accessed May 31, 2014).

58. This is a certification program developed by the U.S. Green Building Council. Silver is the second of four categories of certification for buildings. See the website http://www.usgbc.org/leed#overview (accessed May 25, 2014).

59. G. A. Gonzalez, *Urban Sprawl, Global Warming, and the Empire of Capital* (Albany: State University of New York Press, 2009).

60. Interview, research officer, Chesapeake Bay Foundation, June 19, 2008.

61. A. Olivo, "He Helps Others See His Vision of Redeveloped Tysons Corner," *Washington Post,* January 1, 2014.

62. Interview, committee chairman, McLean Citizens Association, May 29, 2008.

63. A. Gardner, "Plan to Remake Tysons Corner Envisions Dense Urban Center," *Washington Post,* May 29, 2008, A1.

64. Interview, president, Tysons Tunnel Campaign, May 22, 2008.

65. J. O'Connell, "Fairfax County Plans 154 Acres of Urban Parks for Tysons," *Washington Post,* April 25, 2014.

66. Personal communication, principal, Synergy/Planning Inc., Warrenton, VA, January 24, 2009. Among the reservations here are the manner in which the superimposition of a grid of streets remains oriented toward the automobile rather than the pedestrian, the extent to which planned building densities around the metro line will deliver the anticipated outcomes, and the more general notion that property development continues to be oriented to short-term exchange rather than longer-term use values.

67. Interview, president and CEO, Fairfax County Chamber of Commerce, April 8, 2008.

68. Mastran, "Tysons Corner, Virginia," 76.

69. Interviews, president and CEO, Fairfax County Chamber of Commerce, April 8, 2008; president, Tysons Tunnel Campaign, May 22, 2008; and executive director, Coalition for Smarter Growth, May 21, 2008.

70. J. M. Zenzen, *Battling for Manassas: The Fifty-Year Preservation Struggle at Manassas National Battlefield Park,* with a foreword by E. C. Bearss (University Park: Pennsylvania State University Press 1998).

71. J. Garreau, "The Last Thing He Expected Was a Fight," *Washington Post,* July 28, 1991, W14.

72. Quoted in C. Reilly and V. Zapan, "Tysons Corner Is Unofficially Dropping the 'Corner' from Its Name," *Washington Post,* October 5, 2012.

73. Interview, executive director, Coalition for Smarter Growth, May 21, 2008.

74. Ceruzzi, *Internet Alley.*

75. Interview, vice president, EDAW Inc., Alexandria, April 3, 2008.

76. Interviews, president, Tysons Tunnel Campaign, May 22, 2008, and former Fairfax County supervisor, May 28, 2008, respectively.

77. Interview, chair, Tysons Corner Land Use Task Force, April 4, 2008.

78. Interviews, Judy Meany, professor, Dominion University, June 19, 2008, and planner, Planning Department, Fairfax County, April 4, 2008.

79. Interview, senior manager, Development, Tysons Corner Center, May 29, 2008.

80. Interview, senior vice president, Development Services, West Group, May 28, 2008.

81. Interview, senior manager, Development, Tysons Corner Center, May 29, 2008.

82. Interview, partner, Davis Carter Scott, May 20, 2008.

83. Interview, senior vice president, Development Services, West Group, May 28, 2008.

84. Interview, partner, The Georgelas Group of Companies, April 22, 2010.

85. Tysons Corner Collaborative, *Taming Tysons: Transforming the Quintessential Edge City,* n.d., http://www.tamingtysons.com/tysons/tth2.pdf (accessed May 26, 2014).

86. Interview, lawyer, Holland & Knight LLP, April 22, 2010.

87. Greater Washington 2050 Coalition, *Region Forward* (Washington, DC, 2010).

88. Interview, chief of housing and planning, Metropolitan Washington Council of Governments, Washington, DC, May 21 2008.

89. Interview, president, Northern Virginia Transportation Alliance, May 29, 2008.

90. Interview, committee chairman, McLean Citizens Association, May 29, 2008.

91. This is the marketing statement accompanying the Tysons Partnership logo.

92. Interview, vice president, EDAW Inc., Alexandria, April 3, 2008.

93. J. Garreau, "The Shadow Governments: More Than 2000 Unelected Units Rule in New Communities," *Washington Post,* June 14, 1987, A1.

94. Interview, former Fairfax County supervisor, May 28, 2008.

95. Interview, president and CEO, Fairfax County Chamber of Commerce, April 8, 2008. Also see Gardner, "Plan to Remake Tysons Corner."

96. Interviews, lawyer, Holland & Knight LLP, April 22, 2010, and member, Tysons Corner Land Use Task Force and Dranesville resident, April 20, 2010.

97. See the web page http://tysonspartnership.org/the-partnership/partnership-councils (accessed May 15, 2014). The Tysons Partnership has a thirty-five-member board of directors sitting for renewable three-year terms. "The Board provides fiduciary oversight to assure legal, ethical and prudent operation. The Board also defines Partnership policy, direction, goals and budgets." See the web page http://tysonspartnership.org/the-partnership/board-of-directors (accessed May 15, 2014). "Members of the Tysons Partnership are dues paying employers, landlords, land owners and developers, retailers, hoteliers and hospitality service providers, civic groups, and professional consultants that are physically located or own property within Tysons. Fairfax County government is also in the Partnership." See the web page http://tysonspartnership.org/the-partnership/members (accessed May 15, 2014).

98. Interview, supervisor, Fairfax County, May 29, 2008. In 1964, as a self-contained newly master-planned city on the edge of the Washington, D.C., metropolitan area, Reston's vision was as "a creative solution to the twin danger [of] unsightly suburban sprawl and haphazard urban sprawl" (Robert Simon, quoted in K. MacDonald, "Reston Revisited," *Landscape* 32 [1994]: 28–33, 28.) A planned downtown did not take shape until the 1990s.

99. L. Mozingo, *Pastoral Capitalism: A History of Suburban Corporate Landscapes* (Cambridge, MA: MIT Press, 2011), 220.

100. E. Dunham-Jones and J. Williamson, *Retrofitting Suburbia: Urban Design Solutions for Redesigning Suburbs* (Chichester: John Wiley & Sons, 2009),191.

101. See the web page http://www.fairfaxcounty.gov/tysons (accessed May 15, 2014).

102. Indeed, despite its youth as a developed location some sites have already been developed and redeveloped at least three times. Interview, Chief, Policy and Plan Development Branch, Planning Division, Fairfax County Planning Department, April 19, 2010.

103. J. O'Connell, "Tysons: The Building of an American City," *Washington Post,* September 24, 2011.

104. Lang, *Edgeless Cities.*

Chapter 7: Schaumburg

1. Schaumburg Outer Planets Corporation, *The Outer Planets: A Regional Master Concept Planned Unit Development* (Arlington Heights, IL, 1973).

2. R. E. Lang, *Edgeless Cities: Exploring the Elusive Metropolis* (Washington, DC: Brookings Institution Press, 2003).

3. W. J. Holderfield, *Schaumburg's Woodfield Mall* (Chicago: Arcadia, 2007), 73.

4. C. Lemus, *The Village of Hoffman Estates: An Atypical Suburb* (Charleston, NC: History Press, 2009), 70–71.

5. Jack Hoffman was Jewish, and sadly, this as much as anything else was indicated to have figured prominently in the reaction of Schaumburg farmers. Interview, former mayor, Hoffman Estates, 11 August 2011.

6. C. O. Shepherd, "Hoffman Estates: A Village Battles over Growth Problems," *Chicago's American*, May 18, 1965, 28.

7. C. Spirou, "Both Center and Periphery: Chicago's Metropolitan Expansion and the New Downtowns," in *The City Revisited: Urban Theory from Chicago, Los Angeles, and New York*, ed. D. R. Judd and D. Simpson (Minneapolis: University of Minnesota Press, 2011), 273–301, 297.

8. Interview, Mayor of Schaumburg, 7 April 2011.

9. Village of Schaumburg, *Community Profile* (Schaumburg, IL, 1980), 1.

10. See J. F. MacDonald and P. J. Prather, "Suburban Employment Centers: The Case of Chicago," *Urban Studies* 31 (1994): 201–217.

11. Village of Schaumburg, *Schaumburg Comprehensive Plan* (Schaumburg, IL, 1996), 117. Schaumburg has been aggressive in marketing itself as an edge city and a prominent and repeat host of annual meetings of edge cities in the United States.

12. Interviews, economic development coordinator, Village of Schaumburg, July 18, 2011 and principal, Medusa Consulting, April 8, 2011.

13. Interview, senior vice president, Office Brokerage, CBRE, April 26, 2011.

14. Interview, economic development coordinator, Village of Schaumburg, July 18, 2011.

15. It would be another ten years or so before some of the first interest was shown in developing land in the regional commercial center, during which time legend has it that Atcher personally persuaded the remaining farmers to wait and sell their land at the right time in order to receive the best value from developers, often arranging a partial exchange for land elsewhere for those who wished to continue farming. Interviews, mayor, Village of Schaumburg, July 13, 2011, and attorney, Holland and Knight LLP, July 15, 2011.

16. A copy of this version can be consulted in the mayor's office in the Village of Schaumburg.

17. Quoted in M. Reifschneider, "Planet Project Is Approved," *Daily Herald*, September 15, 1968.

18. One could be forgiven for mistaking this plan for a major parcel of land in Schaumburg as a copy of that for downtown Seattle as it contained a monorail and a sky needle, along with a sixty-five-story motor lodge, what would at the time have been the world's tallest office building, and a series of residential blocks. Then again, Romano had recruited the architects responsible for those same developments in Seattle.

19. Quoted in M. Reifschneider, "Tall Talk in Schaumburg," *Daily Herald*, July 12, 1968.

20. Interview, attorney, Holland and Knight LLP, July 15, 2011.

21. See N. A. Phelps, A. M. Wood, and D. C. Valler, "A Postsuburban World? An Outline of a Research Agenda," *Environment and Planning A* 42 (2010): 366–383; N. A. Phelps and A. M. Wood, "The New Post-suburban Politics?," *Urban Studies* 48 (2011): 2591–2610; N. A. Phelps, "The Growth Machine Stops? Urban Politics and the Making and Re-making of an Edge City," *Urban Affairs Review* 84 (2012): 670–700.

22. David K. Hamilton, *Governing Metropolitan Areas: Growth and Change in a Networked Age*, 2nd ed. (New York: Routledge, 2014). The fragmented character of the Chicago metropolitan area is also partly a result of a state constitution that saw special-purpose governments proliferate, until the government's reform in 1970. L. Bennett, "Regionalism in a Historically Divided Metropolis," in Koval et al., *The New Chicago*, 277–285, 278.

23. Ibid., 277.

24. B. Lindstrom, "The Metropolitan Mayors Caucus: Institution Building in a Politically Fragmented Metropolitan Region," *Urban Affairs Review* 46 (2010): 37–67, 50. Indeed, as Lindstrom goes on to document, these same areas of initial collaboration have also proved to be significant fissures along which fragile consensus has broken down.

25. Bennett, "Regionalism in a Historically Divided Metropolis," 278.

26. D. Judd, "Theorizing the City," in *The City Revisited: Urban Theory from Chicago, Los Angeles, and New York*, ed. D. R. Judd and D. Simpson (Minneapolis: University of Minnesota Press, 2011), 3–20, 13. See also Bennett, "Regionalism in a Historically Divided Metropolis," 278.

27. R. Bruegmann, "Schaumburg, Oak Brook, Rosemont, and the Recentering of the Chicago Metropolitan Area," in *Chicago Architecture and Design, 1923–1993: Reconfiguration of an American Metropolis*, ed. J. Zukowsky (Chicago: Art Institute of Chicago; Munich: Prestel, 1993), 177.

28. K. Fidel, "The Emergent Suburban Landscape," in Koval et al., *The New Chicago,* 77–81, 77.

29. Interview, former Parks Board and Zoning Board chair, Hoffman Estates, July 18, 2011.

30. Interview, president, Village of Barrington, Illinois, July 19, 2011.

31. B. Lindstrom, "Regional Cooperation and Sustainable Growth: No One Councils of Government in Northeast Illinois," *Journal of Urban Affairs* 20 (1998): 327–342 . Though these COGs have not been linked institutionally to a multipurpose regional body Lindstrom (p. 328) speculated that "these subregional councils represent the revitalization of councils of government as viable institutional arrangements facilitating intergovernmental cooperation for systemwide problems (transportation infrastructure and solid waste management, for example), membership services (e.g., joint purchasing agreements), and legislative lobbying. . . . Their emergence as agents articulating subregional development strategies in the past decade offers one example of a new institutional arrangement that has the possibility to bypass partisan gridlock and find regional solutions in the Chicago metropolitan region."

32. Bennett, "Regionalism in a Historically Divided Metropolis," 284.

33. Ibid., 279.

34. Interview, mayor of Hoffman Estates, July 19, 2011.

35. Interview, executive director, Metropolitan Mayors Caucus, Chicago, July 12, 2011.

36. Lindstrom, "The Metropolitan Mayors Caucus," 40.

37. Judd, "Theorizing the City," 14; Lindstrom, "The Metropolitan Mayors Caucus"; interviews, executive director, Metropolitan Mayors Caucus, July 12, 2011, and economic development coordinator, Village of Schaumburg, April 7, 2011.

38. L. Bennett, "Chicago's New Politics of Growth," in Koval et al., *The New Chicago,* 44–55, 53.

39. Hoffman Herald, "Hoffman Board Votes to Merge," *Hoffman Herald,* September 1, 1960, 1.

40. Quoted in Hernandez, "Atcher Wouldn't 'Admit' Hoffman Estates," *Schaumburg Pioneer,* November 4, 1993, 7–8. Robert Atcher's animosity in this case was doubtless partly fueled by his own pride in the building of the complete community of Schaumburg, but also by concerns over Hoffman Estates' reputation. Former mayor Virginia Hayter later admitted that "Hoffman Estates' reputation at the time was "everything is for sale."' Lemus, *The Village of Hoffman Estates,* 100.

41. Lemus, *The Village of Hoffman Estates,* 117.

42. Interview, government relations officer, Metropolitan Planning Council, Chicago, July 20, 2011.

43. Metropolitan Planning Council, *Regional Tax Policy Task Force: Report to the CMAP Board* (Chicago, 2012), 10.

44. Ibid., 9.

45. Interview, mayor, Hoffman Estates, July 19, 2011.

46. Interview, government relations officer, Metropolitan Planning Council, Chicago, July 20, 2011.

47. Interview, program director, Transportation and Community Development, Community Neighborhood Trust, Chicago, July 20, 2011.

48. Lindstrom, "Regional Cooperation and Sustainable Growth," 336.

49. Interview, secretary, Northwest Council of Governments, July 20, 2011.

50. D. K. Hamilton, "Regimes and Regional Governance: The Case of Chicago," *Journal of Urban Affairs* 24 (2002): 203–223; J. Schwieterman, "Coalition Politics and America's Premier Transportation Hub," in Koval et al., *The New Chicago,* 288–294.

51. Interview, government relations officer, Metropolitan Planning Council, Chicago, July 20, 2011.

52. Interview, former Parks Board and Zoning Board chair, Hoffman Estates, July 18, 2011.

53. Interview, vice president, Chicago Metropolis 2020, April 11, 2011.

54. Interview, executive officer, Northwest Council of Governments, July 20, 2011.

55. Interview, mayor, Hoffman Estates, July 19, 2011.

56. J. Cidell, "Fear of a Foreign Railroad: Transnationalism, Trainspace, and (Im) mobility in the Chicago Suburbs," *Transactions of the Institute of British Geographers* 37 (2012): 593–608.

57. Metropolitan Planning Council, *Rolling Meadows: Preserving Local Housing Options in the Path of Redevelopment* (Chicago, n.d.), http://www.metroplanning.org (accessed April 23, 2014).

58. Ibid., 8.

59. See the web page http://vhiis2.ci.schaumburg.il.us/publicdocs/Cracker%20Barrel%20Archive/Winter%202012%20Cracker%20Barrel%20Flip/index.html#/0 (accessed April 14, 2015). The same report mentions a new plan for a tax increment financing (TIF) district being drawn up. See also http://www.ci.schaumburg.il

.us/Permit/PLicense/Documents/TIF%20Eligibility%20Study.pdf (accessed May 25, 2014).

60. Interviews, government relations officer, Metropolitan Planning Council, Chicago, July 20, 2011 and director, Community Development, Village of Schaumburg, July 18, 2011.

61. Teska Associates, *North Schaumburg Concept Plan: Village of Schaumburg, Illinois* (Draft of October 30) (Schaumburg: Village of Schaumburg, 2013), 1.

62. Ibid.

63. E. Dunham-Jones and J. Williamson, *Retrofitting Suburbia: Urban Design Solutions for Redesigning Suburbs* (Chichester: John Wiley & Sons, 2009). The authors describe how "isolated, privately owned shopping malls and aging office parks surrounded by asphalt are being torn down and replaced with multiblock, mixed-use town centers, many with public squares and greens" (vi) and note that "the most dramatic and prevalent retrofits tend to be on dead mall sites" (4).

64. Interview, former mayor, Hoffman Estates, August 11, 2011.

65. Holderfield, *Schaumburg's Woodfield Mall*, 81.

66. Interview, manager, Woodfield Mall, July 15, 2011.

67. Interview, senior manager, Long Range Planning, Pace Bus, Arlington Heights, July 13, 2011.

68. L. Mozingo, *Pastoral Capitalism: A History of Suburban Corporate Landscapes* (Cambridge, MA: MIT Press, 2011).

69. See the web page http://www.ci.schaumburg.il.us/Permit/PLicense/Documents/TIF%20Eligibility%20Study.pdf (accessed May 25, 2014).

70. Chicago Metropolitan Area Planning Commission, *GOTO 2040: Comprehensive Regional Plan* (Chicago: CMAP, 2010), 66.

71. Interview, executive director, Metropolitan Mayors Caucus, Chicago, July 12, 2011.

72. Interview, program director Transportation and Community Development, Center for Neighborhood Technology, Chicago, July 20, 2011.

73. Interview, mayor, Arlington Heights, April 11, 2011.

74. Holderfield, *Schaumburg's Woodfield Mall*, 86.

75. Ibid.

76. Interview, mayor, Hoffman Estates, July 19, 2011.

77. Interview, attorney, Holland and Knight LLP, July 15, 2011.

78. Interview, manager, Woodfield Mall, July 15, 2011.

79. Lindstrom, "Regional Cooperation and Sustainable Growth," 331.

80. Interview, president, Village of Barrington, July 19, 2011.

81. See the website http://www.nwpa.us (accessed May 31, 2014).

82. Interview, program director, Transportation and Community Development, Center for Neighborhood Technology, July 20, 2011.

83. Interview, senior manager, Long Range Planning, Pace Bus, Arlington Heights, July 13, 2011.

84. Interview, senior manager, Long Range Planning, Pace Bus, Arlington Heights, July 13, 2011.

85. B. Scheer and M. Petkov, "Edge City Morphology: A Comparison of Commercial Centers," *Journal of the American Planning Association* 64 (1998): 298–310.

86. K. Manson, "Airport Expansion Foes Refocus Efforts," *Chicago Tribune*, April 7, 2010.

87. Interview, executive officer, Northwest Municipal Conference, July 20, 2011.

88. Lang, *Edgeless Cities*.

89. Scheer and Petkov, "Edge City Morphology," 308.

Chapter 8: Conclusion

1. A. Forsyth, "Defining Suburbs," *Journal of Planning Literature* 27 (2012): 270–281, 270.

2. A. Althubaity and A. E. G. Jonas, "Suburban Entrepreneurialism: Redevelopment Regimes and Co-ordinating Metropolitan Development in Southern California," in *The Entrepreneurial City: Geographies of Politics, Regime, and Representation*, ed. T. Hall and P. Hubbard (Chichester: John Wiley & Sons, 1998), 149–172, 150.

3. R. Harris, "Meaningful Types in a World of Suburbs," in *Suburbanization in Global Society*, ed. M. Clapson and R. Hutchinson (Bingley, UK: Emerald, 2010), 15–50.

4. E. Dunham-Jones and J. Williamson, *Retrofitting Suburbia: Urban Design Solutions for Redesigning Suburbs* (Chichester: John Wiley & Sons, 2009), 18.

5. U. Beck, *Risk Society: Toward a New Modernity* (London: Sage, 1992); K. Dennis and J. Urry, *After the Car* (Cambridge: Polity Press, 2009).

6. E. Soja, *Postmetropolis* (Oxford: Wiley-Blackwell, 2000); M. Dear and N. Dahmann, "Urban Politics and the Los Angeles School of Urbanism," *Urban Affairs Review* 44 (2008): 266–279.

7. See, for example, A. Amin, ed., *Post-Fordism: A Reader* (Oxford: Blackwell, 1992).

8. F. J. Turner, *The Frontier in American History* (New York: Holt, Rinehart and Winston, 1962 [1920]), 2.

9. D. Hayden, *Building Suburbia: Green Fields and Urban Growth, 1820–2000* (New York: Vintage, 2004).

10. See N. A. Phelps, A. M. Wood, and D. C. Valler, "A Postsuburban World? An Outline of a Research Agenda," *Environment and Planning A* 42 (2010): 366–383.

11. T. S. Chapin, "From Growth Controls to Comprehensive Planning to Smart Growth: Planning's Emerging Fourth Wave," *Journal of the American Planning Association* 78 (2012): 5–15.

12. Beck, *Risk Society*.

13. Mark Clapson, *Suburban Century: Social Change and Urban Growth in England and the USA* (Oxford: Berg, 2003), 2.

14. Compare, for example, J. K. De Jong, *New SubUrbanisms* (London: Routledge, 2014), and E. Eidlin, "The Worst of All Worlds: Los Angeles, California, and the Emerging Reality of Dense Sprawl," *Transportation Research Record* 1092 (2005): 1–9.

15. N. Smith, *Uneven Development: Nature, Capital, and the Production of Space* (Oxford: Basil Blackwell, 1979).

16. Not the least because, for example, any post-automobile era would be constituted from the likes not only of policy elements such as deprivatization and post-car transport policies but also of wider developments such as the emergence of new fuel systems, new materials, smart cars, digitization, new work/living patterns and disruptive innovation. Dennis and Urry, *After the Car*.

17. J. M. Sellers, "Re-Placing the Nation: An Agenda for Comparative Urban Politics," *Urban Affairs Review* 40 (2005): 419–445, 441.

18. R. Bruegmann, *Sprawl: A Brief History* (Chicago: University of Chicago Press, 2005).

19. A. Duany, E. Plater-Zyberk, and J. Speck, *Suburban Nation: The Rise of Sprawl and the Decline of the American Dream* (New York: North Point Press, 2000), 13.

20. P. Filion, "Optimistic and Pessimistic Perspectives on the Evolution of the North American Suburbs," *Planning Theory and Practice* 14 (2013): 411–413.

21. H. B. Franklin, *The Future Perfect: American Science Fiction of the Nineteenth Century* (New York: Oxford University Press, 1962), quoted in J. Gold, *The Experience of Modernism: Modern Architects and Urban Transformation, 1954–72* (London: Spon, 1997), 210.

22. P. L. Knox, *Metroburbia USA* (New Brunswick, NJ: Rutgers University Press, 2009).

23. R. E. Lang, and J. B. LeFurgy, *Boomburbs: The Rise of America's Accidental Cities* (Washington, DC: Brookings Institution Press, 2007).

24. N. A. Phelps, N. Parsons, D. Ballas, and A. Dowling, *Post-Suburban Europe: Planning and Politics at the Margins of Europe's Capital Cities* (Basingstoke: Palgrave Macmillan, 2006); N. A. Phelps, A. Tarazona Vento, and S. Roitman, "The Suburban Question: Grassroots Politics and Place Making in Spanish Suburbs," *Environment and Planning C* (2015).

25. S. P. Erie and S. A. MacKenzie, "From the Chicago School to the L.A. School: Whither the Local State?," in *The City Revisited: Urban Theory from Chicago, Los Angeles, and New York*, ed. D. R. Judd and D. Simpson (Minneapolis: University of Minnesota Press, 2011), 104–134.

26. On grassroots politics in the potential remaking of Los Angeles as a livable and just city, see R. Gottlieb, M. Vallianatos, R. M. Freeer, and P. Dreier, *The Next Los Angeles: The Struggle for a Livable City* (Berkeley: University of California Press, 2006).

27. See S. Hirt, "Home, Sweet Home: American Residential Zoning in Comparative Perspective," *Journal of Planning Education and Research* 33 (2013): 292–309; and R. A. Beauregard, *When America Became Suburban* (Minneapolis: University of Minnosota Press, 2006).

28. G. A. Gonzalez, *Urban Sprawl, Global Warming, and the Empire of Capital* (Albany, NY: SUNY Press, 2009), 112.

29. See P. G. Newman and J. R. Kenworthy, "'Peak Car Use': Understanding the Demise of Automobile Dependence," *World Transport Policy and Practice* 17 (2011): 31–42; and P. G. Newman and J. R. Kenworthy, "Gasoline Consumption and Cities," *Journal of the American Planning Association* 55 (1989): 24–37.

30. J. L. Grant, "Suburbs in Transition," *Planning Theory and Practice* 14 (2013): 391–392, 391.

31. A. Forsyth, "Suburbs in Global Context: The Challenges of Continued Growth and Retrofitting," *Planning Theory and Practice* 14 (2013): 403–406, 405.

32. F. Wu and N. A. Phelps, "(Post) Suburban Development and State Entrepreneurialism in Beijing's Outer Suburbs," *Environment and Planning A* 43 (2011): 410–430.

33. A. Roy, "The 21st-Century Metropolis: New Geographies of Theory," *Regional Studies* 43 (2009): 819–830, 826.

34. J. Glassman, "Primitive Accumulation, Accumulation by Dispossession, Accumulation by 'Extra-economic' Means," *Progress in Human Geography* 30 (2006): 608–625.

35. D. T. Rodgers, *Atlantic Crossings* (Cambridge, MA: Belknap Press of Harvard University Press, 1998).

36. J. Robinson, *Ordinary Cities: Between Modernity and Development* (London: Routledge, 2006). See also N. A. Phelps, T. Bunnell, M. Miller, and J. Taylor, "Urban Inter-referencing within and beyond a Decentralised Indonesia," *Cities* 39 (2014): 37–49.

Index

Urban and Industrial Environments

Series editor: Robert Gottlieb, Henry R. Luce Professor of Urban and Environmental Policy, Occidental College

Maureen Smith, *The U.S. Paper Industry and Sustainable Production: An Argument for Restructuring*

Keith Pezzoli, *Human Settlements and Planning for Ecological Sustainability: The Case of Mexico City*

Sarah Hammond Creighton, *Greening the Ivory Tower: Improving the Environmental Track Record of Universities, Colleges, and Other Institutions*

Jan Mazurek, *Making Microchips: Policy, Globalization, and Economic Restructuring in the Semiconductor Industry*

William A. Shutkin, *The Land That Could Be: Environmentalism and Democracy in the Twenty-First Century*

Richard Hofrichter, ed., *Reclaiming the Environmental Debate: The Politics of Health in a Toxic Culture*

Robert Gottlieb, *Environmentalism Unbound: Exploring New Pathways for Change*

Kenneth Geiser, *Materials Matter: Toward a Sustainable Materials Policy*

Thomas D. Beamish, *Silent Spill: The Organization of an Industrial Crisis*

Matthew Gandy, *Concrete and Clay: Reworking Nature in New York City*

David Naguib Pellow, *Garbage Wars: The Struggle for Environmental Justice in Chicago*

Julian Agyeman, Robert D. Bullard, and Bob Evans, eds., *Just Sustainabilities: Development in an Unequal World*

Barbara L. Allen, *Uneasy Alchemy: Citizens and Experts in Louisiana's Chemical Corridor Disputes*

Dara O'Rourke, *Community-Driven Regulation: Balancing Development and the Environment in Vietnam*

Brian K. Obach, *Labor and the Environmental Movement: The Quest for Common Ground*

Peggy F. Barlett and Geoffrey W. Chase, eds., *Sustainability on Campus: Stories and Strategies for Change*

Steve Lerner, *Diamond: A Struggle for Environmental Justice in Louisiana's Chemical Corridor*

Jason Corburn, *Street Science: Community Knowledge and Environmental Health Justice*

Peggy F. Barlett, ed., *Urban Place: Reconnecting with the Natural World*

David Naguib Pellow and Robert J. Brulle, eds.,

Power, Justice, and the Environment: A Critical Appraisal of the Environmental Justice Movement

Eran Ben-Joseph, *The Code of the City: Standards and the Hidden Language of Place Making*

Nancy J. Myers and Carolyn Raffensperger, eds., *Precautionary Tools for Reshaping Environmental Policy*

Kelly Sims Gallagher, *China Shifts Gears: Automakers, Oil, Pollution, and Development*

Kerry H. Whiteside, *Precautionary Politics: Principle and Practice in Confronting Environmental Risk*

Ronald Sandler and Phaedra C. Pezzullo, eds., *Environmental Justice and Environmentalism: The Social Justice Challenge to the Environmental Movement*

Julie Sze, *Noxious New York: The Racial Politics of Urban Health and Environmental*

Justice

Robert D. Bullard, ed., *Growing Smarter: Achieving Livable Communities, Environmental Justice, and Regional Equity*

Ann Rappaport and Sarah Hammond Creighton, *Degrees That Matter: Climate Change and the University*

Michael Egan, *Barry Commoner and the Science of Survival: The Remaking of American Environmentalism*

David J. Hess, *Alternative Pathways in Science and Industry: Activism, Innovation, and the Environment in an Era of Globalization*

Peter F. Cannavò, *The Working Landscape: Founding, Preservation, and the Politics of Place*

Paul Stanton Kibel, ed., *Rivertown: Rethinking Urban Rivers*

Kevin P. Gallagher and Lyuba Zarsky, *The Enclave Economy: Foreign Investment and Sustainable Development in Mexico's Silicon Valley*

David N. Pellow, *Resisting Global Toxics: Transnational Movements for Environmental Justice*

Robert Gottlieb, *Reinventing Los Angeles: Nature and Community in the Global City*

David V. Carruthers, ed., *Environmental Justice in Latin America: Problems, Promise, and Practice*

Tom Angotti, *New York for Sale: Community Planning Confronts Global Real Estate*

Paloma Pavel, ed., *Breakthrough Communities: Sustainability and Justice in the Next American Metropolis*

Anastasia Loukaitou-Sideris and Renia Ehrenfeucht, *Sidewalks: Conflict and Negotiation over Public Space*

David J. Hess, *Localist Movements in a Global Economy: Sustainability, Justice, and Urban Development in the United States*

Julian Agyeman and Yelena Ogneva-Himmelberger, eds., *Environmental Justice and Sustainability in the Former Soviet Union*

Jason Corburn, *Toward the Healthy City: People, Places, and the Politics of Urban Planning*

JoAnn Carmin and Julian Agyeman, eds.,*Environmental Inequalities Beyond Borders: Local Perspectives on Global Injustices*

Louise Mozingo, *Pastoral Capitalism: A History of Suburban Corporate Landscapes*

Gwen Ottinger and Benjamin Cohen, eds., *Technoscience and Environmental Justice: Expert Cultures in a Grassroots Movement*

Samantha MacBride, *Recycling Reconsidered: The Present Failure and Future Promise of Environmental Action in the United States*

Andrew Karvonen, *Politics of Urban Runoff: Nature, Technology, and the Sustainable City*

Daniel Schneider, *Hybrid Nature: Sewage Treatment and the Contradictions of the Industrial Ecosystem*

Catherine Tumber, *Small, Gritty, and Green: The Promise of America's Smaller Industrial Cities in a Low-Carbon World*

Sam Bass Warner and Andrew H. Whittemore, *American Urban Form: A Representative History*

John Pucher and Ralph Buehler, eds., *City Cycling*

Stephanie Foote and Elizabeth Mazzolini, eds., *Histories of the Dustheap: Waste, Material Cultures, Social Justice*

David J. Hess, *Good Green Jobs in a Global Economy: Making and Keeping New Industries in the United States*

Joseph F. C. DiMento and Clifford Ellis, *Changing Lanes: Visions and Histories of Urban Freeways*

Joanna Robinson, *Contested Water: The Struggle Against Water Privatization in the United States and Canada*

William B. Meyer, *The Environmental Advantages of Cities: Countering Commonsense Antiurbanism*

Rebecca L. Henn and Andrew J. Hoffman, eds., *Constructing Green: The Social Structures of Sustainability*

Peggy F. Barlett and Geoffrey W. Chase, eds., *Sustainability in Higher Education: Stories and Strategies for Transformation*

Isabelle Anguelovski, *Neighborhood as Refuge: Community Reconstruction, Place-Remaking, and Environmental Justice in the City*

Kelly Sims Gallagher, *The Global Diffusion of Clean Energy Technology: Lessons from China*

Vinit Mukhija and Anastasia Loukaitou-Sideris, eds., *The Informal City: Settings, Strategies, Responses*

Roxanne Warren, *Rail and the City: Shrinking Our Carbon Footprint and Reimagining Urban Space*

Marianne Krasny and Keith Tidball, *Civic Ecology: Adaptation and Transformation from the Ground Up*

Julian Agyeman and Duncan McLaren, *Sharing Cities: Enhancing Equity, Rebuilding Community, and Cutting Resource Use*

Jessica Smartt Gullion, *(In)Visibility in the Gas Field: Health, Activism, and the Barnett Shale*

Nicholas A. Phelps, *Sequel to Suburbia: Glimpses of America's Post-Suburban Future*